The Complete Guide to the ABC Molecular Biology Certification Exam

The Complete Guide to the ABC Molecular Biology Certification Exam

Tiffany Roy and Tatum Price

CRC Press is an imprint of the
Taylor & Francis Group, an **informa** business

CRC Press
Taylor & Francis Group
6000 Broken Sound Parkway NW, Suite 300
Boca Raton, FL 33487-2742

© 2020 by Taylor & Francis Group, LLC
CRC Press is an imprint of Taylor & Francis Group, an Informa business

No claim to original U.S. Government works

International Standard Book Number-13: 978-0-367-82133-3 (Hardback)
International Standard Book Number-13: 978-1-4987-5392-0 (Paperback)
International Standard Book Number-13: 978-0-429-27153-3 (e-Book)

This book contains information obtained from authentic and highly regarded sources. Reasonable efforts have been made to publish reliable data and information, but the author and publisher cannot assume responsibility for the validity of all materials or the consequences of their use. The authors and publishers have attempted to trace the copyright holders of all material reproduced in this publication and apologize to copyright holders if permission to publish in this form has not been obtained. If any copyright material has not been acknowledged please write and let us know so we may rectify in any future reprint.

Except as permitted under U.S. Copyright Law, no part of this book may be reprinted, reproduced, transmitted, or utilized in any form by any electronic, mechanical, or other means, now known or hereafter invented, including photocopying, microfilming, and recording, or in any information storage or retrieval system, without written permission from the publishers.

For permission to photocopy or use material electronically from this work, please access www.copyright.com (http://www.copyright.com/) or contact the Copyright Clearance Center, Inc. (CCC), 222 Rosewood Drive, Danvers, MA 01923, 978-750-8400. CCC is a not-for-profit organization that provides licenses and registration for a variety of users. For organizations that have been granted a photocopy license by the CCC, a separate system of payment has been arranged.

Trademark Notice: Product or corporate names may be trademarks or registered trademarks, and are used only for identification and explanation without intent to infringe.

Visit the Taylor & Francis Web site at
http://www.taylorandfrancis.com

and the CRC Press Web site at
http://www.crcpress.com

Contents

Purpose of This Book	**xi**
Authors	**xiii**

1 General Knowledge Introduction — 1

1.1	History	1
	1.1.1 Significant Figures	2
	1.1.2 Evolution of the Practice	5
1.2	Crime Scene Preservation	8
1.3	Evidence Handling	14
	1.3.1 Evidence Recognition and Collection	14
	1.3.2 Evidence Packaging and Preservation	17
	1.3.3 Evidence Classes (Class/Individual)	19
1.4	Crime Laboratory Operations—Overview	19
	1.4.1 Forensic Biology	20
	1.4.2 Controlled Substances	20
	1.4.3 Toxicology	29
	1.4.4 Trace Analysis	29
	1.4.5 Latent Fingerprints	29
	1.4.6 Questioned Documents	33
	1.4.7 Fire Debris	33
	1.4.8 Firearms and Toolmarks	34
	1.4.9 Digital Evidence	36
	Notes	37

2 Quality Assurance and Quality Control — 39

2.1	Quality Assurance and Quality Control	39
	2.1.1 Accreditation	39
	2.1.2 Accrediting Bodies	40
	2.1.3 Personnel Certification	41
	2.1.4 Standardization	50
2.2	Quality Assurance/Quality Control Application	51
2.3	Document/Data Management	56
2.4	Safety	56
	2.4.1 Chemical Hygiene	56

2.4.2 Universal Precautions	58
2.4.3 Hazardous Waste	58
Bibliography	59

3 Basic Legal and Scientific Concepts 61

3.1 Legal Decisions	61
3.2 Legal Terms	62
3.3 Court Testimony	63
3.4 Procedural Law	64
3.5 General Science Terms and Principles	65
3.5.1 General Chemistry Concepts	66
3.5.2 General Biology Concepts	71
3.5.3 General Physics Concepts	72
3.5.4 General Physiology Concepts	73
3.5.5 General Statistics Concepts	73
3.5.6 Logic	75
Notes	77
Bibliography	77

4 Principles and Concepts of Biological Screening 79

4.1 Biological Screening Tests	79
4.1.1 Blood	79
4.1.1.1 Blood Typing	80
4.1.1.2 Presumptive Tests for Blood	83
4.1.1.3 Confirmatory Tests for Blood	84
4.1.1.4 Blood Species Identification	87
4.1.2 Semen	88
4.1.2.1 Components of Semen	88
4.1.2.2 Presumptive Tests for Semen	89
4.1.2.3 Confirmatory Tests for Semen	91
4.1.3 Saliva	93
4.1.3.1 Presumptive Tests for Saliva	93
4.1.3.2 Confirmatory Tests for Saliva	94
4.1.4 Urine	95
4.1.4.1 Properties of Urine	95
4.1.4.2 Presumptive Tests for Urine	96
4.1.4.3 Confirmatory Tests for Urine	96
4.1.5 Feces	97
Bibliography	97

5 Anatomy and Cell Biology   99

5.1   Anatomy, Physiology, Reproductive Biology   99
    5.1.1   Biochemistry of Physiological Fluids   99
5.2   Cellular and Molecular Biology   99
    5.2.1   Cell Morphology   99
    5.2.2   Cells and Chromosomes   103
    5.2.3   Chromosomal Organization   107
    5.2.4   Cellular DNA Content   107
    5.2.5   Cell Division   108
    5.2.6   DNA Structure   109
    5.2.7   Transcription and Translation   111
    5.2.8   Replication   114
        5.2.8.1   DNA Organization   114
        5.2.8.2   Replication Forks and Bubbles   114
        5.2.8.3   Enzymes Involved in DNA Replication   115
        5.2.8.4   Proofreading Mechanisms   116
    5.2.9   Mutation Mechanisms and Rates   118
        5.2.9.1   Kinds of Mutations   118
        5.2.9.2   Repair   119
Bibliography   121

6 Concepts in Genetics, Biochemistry, and Statistics   123

6.1   Genetics   123
    6.1.1   Mendelian (Autosomal) Genetics   124
        6.1.1.1   Rules of Inheritance   126
        6.1.1.2   Human Pedigrees   127
    6.1.2   Non-Mendelian   130
        6.1.2.1   Y-Chromosomal Inheritance   131
        6.1.2.2   Mitochondrial DNA   133
    6.1.3   Cytogenetics: Chromosomal Abnormalities   137
        6.1.3.1   Nondisjunction: Mitotic and Meiotic   137
        6.1.3.2   Chromosomal Abnormalities   138
    6.1.4   Genetic Disease   138
Note   140
Bibliography   140

viii

Contents

7 Concepts in Genetics, Biochemistry, and Statistics cont.   141

7.1	Population Genetics	141
	7.1.1 Hardy–Weinberg	142
	7.1.2 Mechanisms of Evolution	145
	7.1.2.1 Mutation	145
	7.1.2.2 Selection	145
	7.1.3 Statistics and Probability	146
	7.1.3.1 Likelihood Ratios	146
	7.1.3.2 Pd and Pi	147
	7.1.4 Population Databases	148
7.2	Non-Human Molecular Applications	149
	7.2.1 Animal Forensic DNA Applications	150
	7.2.2 Plant Forensic DNA Applications	150
	7.2.3 Microbial DNA Applications	152
	Bibliography	152

8 History and Standards of DNA Evidence   155

8.1	Types of Evidence	155
	8.1.1 Criminal	156
	8.1.2 Non-Criminal	158
	8.1.3 Missing-Person/Mass Disaster	158
	8.1.4 Kinship	158
	8.1.5 Databanking	160
8.2	Evolution of the Discipline	160
	8.2.1 Antigen and Immunological Systems	161
	8.2.2 Protein and Enzyme Polymorphisms	164
	8.2.3 DNA Polymorphisms	167
	8.2.3.1 RFLP	167
	8.2.3.2 PCR, qPCR, RT-PCR	168
	8.2.3.3 Genetic Markers	178
	8.2.3.4 Automation	181
8.3	Accepted Standards and Practices	183
	8.3.1 Quality Assurance Standards for Forensic DNA Laboratories	183
	8.3.2 Quality Assurance Standards for DNA Databasing Laboratories	183
	8.3.3 SWGDAM	184
	8.3.4 CODIS	184
	8.3.5 Validation for Introduction of New Technologies	184
	Notes	185
	Bibliography	185

Contents

ix

9 DNA Testing Process Part 1

187

9.1 Casework Documentation and Reporting (Results and Conclusions) — 187
    9.1.1 Case Management — 187
    9.1.2 Evaluate Requests for Analysis to Determine Appropriate Evidence Screening and Comparisons to Develop the Most Useful Information — 187
    9.1.3 Establishing Case Record — 188
9.2 Process Analysis — 190
    9.2.1 Considerations of Analytical Limitations — 190
    9.2.2 Required Testing Controls — 190
9.3 Reporting — 191
    9.3.1 Requirements — 191
    9.3.2 Quantitative/Qualitative Conclusions — 192
9.4 Artificial Intelligence — 193
    9.4.1 Second Read Software and Automation — 193
9.5 Visualization Tools/Techniques — 194
    9.5.1 Microscopy — 194
    9.5.2 Electrophoresis — 198
        9.5.2.1 Gel Electrophoresis — 198
        9.5.2.2 Capillary Electrophoresis — 202
    9.5.3 Fluorescence — 203
Notes — 204
Bibliography — 204

10 DNA Testing Process Part 2

205

10.1 Isolation and Purification of Nucleic Acids — 205
    10.1.1 Organic Extraction — 206
    10.1.2 Differential Extraction — 207
    10.1.3 Chelex Extraction — 208
    10.1.4 Silica-Based Extraction — 209
10.2 Quantification — 210
    10.2.1 Slot Blot Assay — 210
    10.2.2 Intercalating Dye Assay — 210
    10.2.3 Quantitative PCR — 211
10.3 Polymerase Chain Reaction — 212
10.4 DNA Typing Technology — 212
    10.4.1 Fragment Analysis/Short Tandem Repeat Analysis — 212
        10.4.1.1 Theory — 212

	10.4.1.2 Application/Processes	213
	10.4.1.3 Interpretation/Results	213
Bibliography		215

Index 217

Purpose of This Book

In 2009, the National Academy of Sciences authored the report, *Strengthening Forensic Science in the United States: A Path Forward.* In it, the Committee expressed the importance of accreditation and certification. Accreditation, long recognized by public labs as an important benchmark in quality, was recognized as an important way to standardize laboratories that provide forensic services. Certification, which plays an important role in many other professions, was suggested as an important method of oversight in the forensic sciences. In many other areas of the professional world, there are resources available to assist those who endeavor to take professional certification exams. Doctors and nurses, lawyers, accountants, and engineers have commercial materials available to them to assist them in their pursuit of professional certification and licensure. Companies like Kaplan and Princeton Review produce audio, video, and print materials to help professionals be successful. No such materials exist for forensic professionals. It has long been required that doctors and lawyers seek certification or licensure, but because it is still not mandatory, for-profit companies may not yet see the financial investment in producing preparation material for forensic certification.

In the ten years since the report was published, there has been increased interest in forensic certification. Many scientists have voluntarily taken part in certification programs driven by a genuine desire to elevate the profession. There are resources that have been developed for certification programs like the International Association for Identification disciplines, which provide certification for field disciplines like footwear impressions and crime scene investigation. Most of these are professional preparation courses provided by forensic professionals who have experienced the preparation process themselves. But for working forensic scientists with families, jobs, and professional obligations, the resources available are not nearly sufficient.

As one of the authors of this text, this problem became apparent to me in my own pursuit of forensic certification. In February 2011, I took and passed the Massachusetts Bar Examination. I spent thousands of dollars on professional preparation materials. I had audio CDs to listen to review material while I drove in my car. I had access to test question banks. I had video lectures, print material, and flash cards. And I was successful. After receiving the bar exam results, I decided the next pursuit would be forensic certification. The two experiences could not have been more different. The American

xi

Board of Criminalistics exam outline and the bibliography that accompanied that were the only guidance available to tailor my study. Because the bar examination is mandatory for lawyers and forensic certification is not, there were no commercial resources. Until certification is required for forensic professionals, it's unlikely that we will see commercial test preparation companies take an interest in forensic certification preparation. It is my hope that this text will help working forensic scientists feel less overwhelmed in their pursuit of forensic certification, as it plays an important role in forensic science quality and standardization.

Tiffany Roy

Authors

TIFFANY ROY, JD is a Forensic DNA expert with over thirteen years of forensic biology experience in both public and private laboratories in the United States. She instructs undergraduates at Palm Beach Atlantic University, University of Maryland University College, and Southern New Hampshire University. She currently acts as a consultant for attorneys and the media in the area of forensic biology through her firm, *ForensicAid, LLC*. Roy holds degrees from Syracuse University, New York, Massachusetts School of Law, and University of Florida in the areas of Biology, Law, and Forensic Science. It is her benchworking forensic DNA experience and personal pursuit of forensic certification that played an influential role in the authorship of this text. In addition to this text, Roy authors two other widely recognized texts in the field of forensic science: *Criminalistics: An Introduction to Forensic Science*, currently in its thirteenth edition, and *Forensic Science: From the Crime Scene to the Crime Lab*, currently in its fourth edition. Her teaching, legal writing, and testimonial experience help her to take complex scientific concepts and make them easily understandable for the non-scientist.

TATUM PRICE, BS is a Forensic Scientist in the Forensic Biology Unit at the Palm Beach County Sheriff's Office. She holds a Bachelor's degree in Forensic Science from Palm Beach Atlantic University, Florida, where she graduated *summa cum laude* with honors. She was a student of Roy's during her undergraduate studies, and she went on to work alongside Roy at *ForensicAid, LLC*. Price provided contributions to the newest editions of the forensic texts *Criminalistics: An Introduction to Forensic Science* as well as *Forensic Science: From the Crime Scene to the Crime Lab*.

General Knowledge Introduction

1

1.1 History

According to the American Board of Criminalistics, a qualified forensic biologist must be able to demonstrate the necessary knowledge, skills, and abilities. The following is a list the Board has published that delineates which knowledge, skills, and abilities one must demonstrate to be successful on the examination for the molecular biology specialty area certification:

- Apply principles of general, cell and molecular biology, biochemistry, genetics, and statistics to the analysis of biological materials.
- Apply chemical, immunological, microscopic, and molecular biological methods in the assessment of unknown biological material.
- Apply specialized techniques to isolate and purify nucleic acids from various biological material.
- Apply specialized techniques to quantify DNA.
- Apply specialized techniques to develop profiles from the polymorphic regions in biological genomes.
- Apply principles of population genetics to then determine the significance of a polymorphic profile in the population.
- Stay abreast of current developments in the field of forensic biology.
- Recognize, collect, secure, and preserve physical evidence.
- Recognize the potential for forensic examinations in areas outside an area of specialization, prioritize the sequence of examinations, and handle evidence accordingly.
- Observe safe practices in the examination of biological evidence.
- Engage in impartial and ethical practices.
- Use and maintain laboratory instrumentation proficiently.
- Evaluate and interpret results of analyses.
- Thoroughly and accurately produce documentation to support results and conclusions.
- Effectively communicate scientific results through written reports.
- Provide sworn testimony regarding analytical methods, techniques, results, and conclusions.
- Employ quality assurance measures to ensure the integrity of the analyses.

- Understand and apply the validations for the introduction of new DNA technologies into the forensic laboratory.
- Be familiar with the documents *Quality Assurance Standards for Forensic DNA Testing Laboratories* and *Quality Assurance Standards for Forensic Databasing Laboratories*.
- Understand uses and practices of the Combined DNA Index System (CODIS).
- Conduct second reads and technical reviews of the analytical work of other forensic biologists.[1]

1.1.1 Significant Figures

To begin to accomplish these goals set forth by the ABC, one must become familiar with the historical figures and evolution of practice, not only in one's subject-matter area, but in forensic science in general. An estimated 60% of the examination questions will focus on general forensic knowledge including topics like history, quality control and assurance, and law. Some of the important figures in the forensic sciences that should be studied for the purposes of the exam are:

Mathieu Orfila

Orfila was a Spanish native widely considered to be the father of forensic toxicology. In 1814, Orfila published the first scientific treatise on the detection of poisons and their effects on animals, establishing forensic toxicology as a legitimate science.[2]

Alphonse Bertillon

Bertillon is known as the father of criminal identification. He established the first scientific means of differentiating and identifying human beings using a series of biometric body measurements. His system, anthropometry, was used for nearly two decades as the first and only means of personal identification before it was replaced by fingerprints.[3]

Introduction

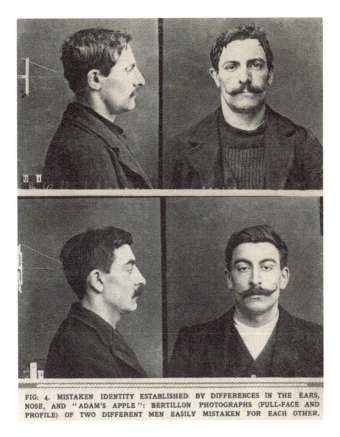

FIG. 4. MISTAKEN IDENTITY ESTABLISHED BY DIFFERENCES IN THE EARS, NOSE, AND "ADAM'S APPLE": BERTILLON PHOTOGRAPHS (FULL-FACE AND PROFILE) OF TWO DIFFERENT MEN EASILY MISTAKEN FOR EACH OTHER.

Figure 1.1 Anthropometry was used to discriminate between individuals based on a series of body measurements. It was widely used until two men (pictured) with the same name and indistinguishable measurements arrived at Fort Leavenworth Prison. (Source: Shutterstock.com)

Hans Gross

Gross was an Austrian prosecutor and judge who authored the first treatise on the application of the sciences to criminal investigation. He introduced the journal *Archiv für Kriminal Anthropologie und Kriminalistik*, which is still used to publish new methods of scientific crime detection.[4]

Edmond Locard

Locard was a Frenchman who is best known for the postulation that every contact leaves a trace. This theory, known as Locard's Exchange Principle, stated that there is a transfer of material between a victim, a suspect, and the crime scene.[5]

Figure 1.2 Locard's exchange principle demonstrates that every time a person comes into contact with another person, place or thing, there is some transfer of material from each. "Every contact leaves a trace."

Francis Galton

Galton was an English statistician and psychologist who wrote *Fingerprints*, the first treatise describing and classifying the unique qualities of fingerprints. This work, written in 1892, laid the foundation for modern-day dactylography.[6]

Leone Lattes

Lattes, a professor of Forensic Medicine at the University of Turin in Italy, developed a procedure for determining the blood group of a dried blood stain in 1915. The test, called the Lattes Crust Test, was groundbreaking because blood-group testing could only be performed on liquid blood prior to its discovery. This test was immediately useful in criminal investigations.[7]

Calvin Goddard

Goddard is recognized as the father of modern firearms analysis. As a U.S. Army colonel, Goddard was the first to use the comparison microscope to

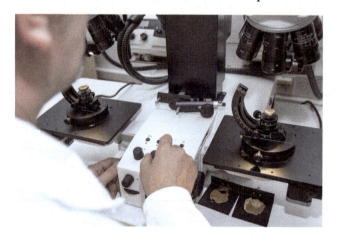

Figure 1.3 Comparison microscopes remain the cornerstone of forensic firearms analysis. (Source: Shutterstock.com)

Introduction

compare bullet markings in order to determine the weapon of origin. From the mid-1920s, Goddard's work established the comparison microscope as an indispensable tool in modern-day ballistic examination.[8]

Albert S. Osborn
Osborn is considered the father of questioned document examination. In 1910, he penned the first significant text in the field, *Questioned Documents*, which is still considered a primary reference in the field today.[9]

Walter C. McCrone
McCrone is considered to be the father of modern microscopy. McCrone, a skilled American chemist and instructor, combined microscopic examination with analytical techniques to characterize various types of evidence. He trained thousands of scientists, wrote hundreds of articles and books, and gave thousands of lectures and presentations on light and electron microscopy over the course of his career.[10]

August Vollmer
Vollmer, a police chief from Berkeley, California, created the first forensic laboratory in the United States at the Los Angeles Police Department. Vollmer started the lab in 1923 and in the 1930s he began the first university institute for criminology and criminalistics at the University of California at Berkeley.[11]

Paul L. Kirk
Kirk is known as one of the most pre-eminent criminalists of modern times. A microscopist and forensic scientist, he was head of the Criminalistics Department at the University of California at Berkeley, where he authored several books in the areas of criminalistics, microscopy, biochemistry, and forensic science. He is best known for his involvement in the bloodstain pattern analysis performed in the Sam Sheppard case.

Karl Landsteiner
Landsteiner was an Austrian-born doctor and chemist who created the modern-day blood-typing classification system using agglutination. His description of the ABO system and Rhesus factor was groundbreaking for the medical field. He was awarded the Nobel Prize for his work in 1930. This discovery allowed for the successful transfusion of blood and transplant of organs, as well as being a means of identifying blood types of donors who may have left blood behind at crime scenes.

1.1.2 Evolution of the Practice

Much of the practice of modern forensic science began in Britain in the mid-to-late 1800s. Criminal investigation was a burgeoning field in that time:

cities were becoming more populous and crime was more prevalent. This increase in crime required more advanced techniques for criminal investigation and crime detection. The identification of the value of fingerprints and the creation of a method for their classification, advances and studies in firearm and bullet markings, forensic pathological examinations, and toxicological tests all paved the way for modern forensic science. Around the turn of the century, American and British investigators began to exchange ideas on the organization of investigative forces and the use of modern techniques for advanced crime fighting and detection. Since then, the practice of forensic science has continually expanded and improved as many joined together to develop the different disciplines of forensics.

Perhaps the most significant advancements in the early stages of forensic biology were the discovery of ABO blood typing and its application to medicine and criminalistics, as well as the ability to perform this type of testing on both fresh liquid and dried bloodstains left at crime scenes. These advancements laid the foundation for blood protein analysis and body fluid identification. These tests were the precursor to modern DNA testing, which is widely regarded as the "gold standard" for forensic analysis.

The basis for DNA analysis was established with the ability to differentiate blood at crime scenes based on type. With the knowledge that there are four main types of blood, some more common and some rarer in the population, it became possible to assign weight when a suspect's blood type matched the type from the crime scene stain. This information was helpful to early forensic scientists and investigators because when the blood types from the crime scene samples were consistent with the blood types from individuals known or believed to be involved in a case, this offered a means for demonstrating a possible link. If a suspect had AB blood, only 3% of the population would be expected to have that blood type. Since fewer members of the population would have that type, this gave weight to the association. The problem faced by many investigators at this time was differentiating blood from a victim and suspect that shared the same blood type, or from multiple suspects who may have shared the same blood type.

Here, advancements in the identification of blood proteins helped to further characterize the blood and allowed for differentiation between individuals who shared the same ABO blood type. One of the first blood proteins that was characterized and used in this way was the Rhesus (Rh) factor. The more proteins that were studied and characterized in the blood, the greater the ability of the investigator to associate that blood with a victim or suspect. In all, more than 100 different blood proteins have been identified and many had common use in forensic science when it came to characterizing bloodstains in forensic testing.

Introduction

The use of blood types and proteins as a means of identification was similar in nature to anthropometry. In anthropometry, the use of several specific physical characteristics helped to identify a person to the exclusion of all others. This method was not unlike a physical description reported by a witness to a crime. In both of these instances, several physical traits are described, and the more observations that are made, the more unique the description becomes. For example, a witness to a crime describes a suspect as being 6' tall with blond hair and blue eyes. Those three criteria alone would rule out a significant portion of the population from fitting that description. With each added criteria (weight, hair length, facial hair, birthmarks, tattoos, etc.), more and more people would be excluded. The goal is to add criteria until a profile exists with such detailed information that only one person in the population would match it exactly.

Blood grouping and protein analysis rely on the same principles of identification. First, the blood type of the sample would be identified. Each blood type has been studied and an estimation of how common or rare it would be for a person to have that type has been calculated by research studies. The blood is then tested to determine whether it is Rh+ or Rh−, which also has an expected frequency in the population. Results of different blood protein tests can be added to the overall profile, each with their own frequency. The frequencies of occurrence are then multiplied together using the product rule. With each new trait, the protein profile of the blood becomes more and more rare, until the chances of a random person also having the identified combination of blood proteins and factors is exceedingly rare. For example, a bloodstain found at a crime scene is determined to be type AB blood. The suspect in the case also has type AB blood, which is only expected to be seen in 3% of people. The blood tests positive for Rhesus factor. Since only 1% of the population would have AB+ blood, the testing of these two criteria has already succeeded in excluding 99% of the population as being a possible contributor of the blood left at the crime scene. Additional factors can and would be tested to further characterize the blood, thereby making it more unique and strengthening the association between the suspected person and the evidence left at the crime scene.

Modern DNA analysis rests upon this foundation as well. Just as in the physical description and the blood type and protein profile, there are traits that are examined in the DNA. Each location on the DNA that is examined in modern-day DNA profiling can vary from person to person. Looking at one location may not give the examiner enough information to make the profile unique, but the more locations examined and the more traits observed, the rarer it becomes.

1.2 Crime Scene Preservation

Preserving the crime scene is of paramount importance. The fidelity of the evidence can easily be lost in the first few minutes or hours of the investigation by failing to properly protect the crime scene and maintain crime scene security. One of the many challenges faced by modern forensic scientists as a result of the popularity of television and movies is increased public interest. When a crime occurs, people are often interested in viewing the scene, watching the officers investigate, and watching the scientists collect evidence. Flashing blue lights and yellow evidence tape attracts onlookers. Making certain that onlookers remain at a safe distance from the scene and evidence will help to ensure the validity of that evidence once it can be analyzed in the lab.

Figure 1.4 The O.J. Simpson case is an example of how poor crime scene preservation methods can detrimentally impact a case and diminish the value of any evidence recovered. (Public domain: retrieved from Wikimedia.)

Introduction

When crimes occur involving multiple victims or people of importance to the community or society, investigative personnel may respond from miles around to assist in the investigation.

Even if most will respond out of genuine desire to offer assistance, allowing too many people into the crime scene risks the destruction of evidence and diminishes the value of the evidence inside. Consider the crime scene where Nicole Brown Simpson and Ronald Goldman were stabbed. Hundreds of police and government personnel responded to that scene to assist in the investigation. The failure of the investigating officers to exclude unnecessary personnel resulted in contamination of the scene by the officers themselves, who left behind footwear impressions and other traces that complicated the analysis of the evidence that existed prior to their arrival. Many hours were wasted by scientific personnel wading through what may have been left behind by the officers and what may have been left behind by the perpetrator. The greater the care an officer or scientist takes to protect and preserve the crime scene, the more faith they will have that the evidence is an accurate representation of what happened during the commission of the crime.

On initial arrival, the priority of the first responder should be preservation of human life. This concern is twofold. If a victim is present who has been injured, every attempt should be made to help that victim. Calls for medical personnel to render aid and bring the injured person or persons to the hospital should be immediate. Once those calls have been made, the first responding officer will take steps to ensure the safety of the scene for him or herself, as well as for the medical personnel and any backup units who were signaled to respond. Preservation of the victim's life and the lives of officers and scientists responding is equally important in the early stages of scene discovery. This requires a thorough search of the premises to ensure no perpetrators remain who could injure those who will be responding to assist with medical care or the impending investigation.

Once the scene has been deemed safe, responding officers should examine it to determine all areas of entry and egress. Evidence may exist in secondary areas in the scene, so all areas where the crimes may have occurred should be identified. Once the scene boundaries have been established, crime scene tape should be used to cordon off the area to limit access. Backup officers arriving should assist in the security effort to ensure that no suspects or members of the public can enter the scene and tamper with evidence or threaten the safety of investigators and scientists present. A member of the investigative team should be posted at the entry point to maintain a list of names of necessary personnel who enter and exit the crime scene area. Every effort should be made to limit the number of people who access the areas where evidence may exist.

Figure 1.5 A designated officer is responsible for ensuring crime scene integrity by keeping a log of when authorized personnel enter and exit. (Source: Shutterstock.com)

It is also extremely important to keep in mind that crimes often involve biological material and human body fluid. These fluids may contain infectious disease which might not be known to the investigator, so it is important for any forensic scientist to treat all crime scenes and the biological material contained in them as infectious. Personal protective equipment should be worn at all times at a crime scene, and this may include long pants, comfortable footwear with closed toes, footwear covers, gloves, lab coats, and face masks. This equipment will prevent harm to the investigator in the form of an infection from exposure to biohazardous material and it will also protect the evidence from contamination by the investigator.

Once the scene is secure and properly isolated from public access, documentation of the scene should occur. There are three main ways to document crime scenes. The first is with handwritten notes and sketches. The first responding officer or scientist will begin taking down information such as: the address of the location where the crime occurred, the names and birthdates of the victims and suspects if they have been identified, the time they received the call for assistance, the date and time of arrival and the date and time of departure, the names of all investigative and scientific personnel present, physical descriptions of the scene, and weather conditions. A sketch or diagram of the scene should be made, including all areas where evidence of the crime may exist. A list of all evidence noted, marked, and collected should exist somewhere in the notes and sketches. Other common details included in the crime scene notes and sketch may include a compass, registrations and serial numbers of vehicles present, and a key that identifies important items of evidence.

Introduction

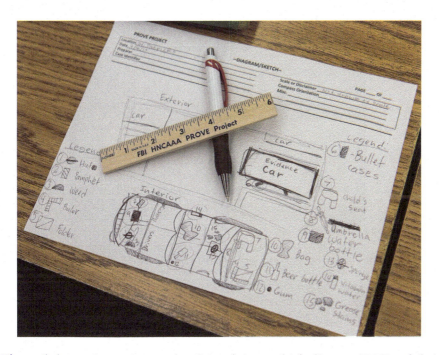

Figure 1.6 Rough crime scene sketch involving a vehicle. (Source: FBI Honolulu.)

There are two general types of sketches for crime scene documentation: rough and finished. A rough sketch is one that is drawn freehand at the scene to give the forensic scientist a general idea of the location and relation of items of evidence at the scene. Several different sketches may be drawn of each area or room, as well as an overall sketch of the scene. Once the scientist is back in the lab, they may wish to create a finished sketch, which may be drawn to scale and may include specifics like measurements of items or stains noted. This can be done with the use of a computer program or with the help of a crime scene template or ruler for neatness and organization.

The crime scene should also be documented with video recording. A walk through of the scene should be done with a video camera to document in real time all the evidence and conditions of the scene. Many video cameras will be equipped with audio, so it's extremely important for those present at the scene to be mindful of the discussions and statements being made in the presence of the videographer. Since the video may be played in the courtroom for the judge or jury to demonstrate what the scene looked like on arrival, those present at the scene should make every effort to either remain silent during the taping or refrain from any discussions that may be interpretive, suggestive, or unprofessional.

Figure 1.7 Photographs of a crime scene are critical to refer to for reconstruction purposes or to examine the original state of the evidence. (Source: Shutterstock.com)

Finally, the scene should be documented with still photography. Photos are necessary to add the close up detail that may not be possible through video recording. Crime scene photos are often taken in a progression from overall shots of the scene as a whole to close up shots of specific items of evidence. The photographer should start in an area where the overall evidence in a room is visible and move closer and closer to areas where there are high concentrations of evidence. Individual close up shots should be taken of items of interest. Photo and video documentation should take place before and after marking items of evidence. When taking photographic evidence at a crime scene, it is also important to remember to include rulers or measuring tape for scale in close up shots of evidentiary items.

Though some search has already taken place to identify the boundaries of the crime scene, a more thorough search of the premises will take place after the initial documentation of the scene to identify any items that may have been concealed by the perpetrator. Depending on the type of crime and the location of the scene, different methods of searching will likely be employed. For example, when investigating an explosion or bombing which may stretch over a large area, a grid search for evidence may be employed. For a scene inside a small bedroom, a spiral search may be sufficient. For searches of large open areas for human remains and evidence of crimes, a line search may be employed.

When dealing with sexual assault crimes, a search of the scene may take place with the naked eye as well as with alternate light source techniques for identifying body fluids. When searching a scene for blood that may have been cleaned in an attempt to conceal the crime, chemical reagents such as Luminol or Bluestar may be necessary to visualize the evidence. A thorough, systematic search based on the unique details of that crime should be employed by the forensic personnel.

Introduction

Figure 1.8 Line search of a large outdoor area. (Source: Getty.)

All items of evidence related to the crime should be documented and listed in the handwritten notes. Each item should be given a unique identifier that should be present in photographs and referenced in notes. There are often placards that are displayed next to each item of evidence at the crime scene. These placards may have number or letter designations, which can be easily referenced in reports and notes. For example, a bloody glove at a crime scene is noted by the investigator. After documenting the item, a placard bearing the number 1 is placed next to the glove, and this can be referenced in subsequent notes and reports as item 1. When the item is collected and packaged, the package will also bear an item number and description on the exterior. This is so that when it is time to examine the items from the scene, the analyst assigned to the case will be able to clearly identify what is contained in each package.

Taking the facts known about the crime and searching based on the characteristics of that crime will help investigators to know what types of evidence may hold probative value for the case. For instance, if officers were investigating a robbery of a convenience store, a thorough search of the surrounding areas may yield evidence discarded by the perpetrators. If the cashier of the convenience store described the suspect as wearing black gloves and a black ski mask, a search of the entrance and egress points yielding those items would immediately be of interest to the forensic investigator. The DNA of the suspect, as well as hairs and fibers from the suspect, may be present on those items. The information from the clerk could be vital in determining the evidentiary value of the items. Another example of when it is important to be able to recognize the value of the evidence uncovered might occur in a missing person case. If a woman disappears from her home and a thorough search of the home using chemical processing uncovers evidence of bloodstains and blood spatter, samples of the stains should be collected so that they can eventually be compared with the DNA profile from the missing woman. In a forensic investigation, there is absolutely no substitute for experience. As a scientist becomes more experienced searching for evidence at crime scenes, this experience will advance his or her ability to recognize evidentiary value.

Once something has been identified as a probative item of evidence, care must be taken to ensure the item won't injure any of the investigators or scientists during subsequent collection, transport, and analysis. Consider an explosives investigation where an incendiary device has been recovered which may contain DNA from the perpetrator. This must first be rendered safe so that it can be collected and transported to the lab for analysis. Rendering an improvised explosive device (IED) safe will likely require the assistance of a trained expert in that area. Guns are frequently identified as evidence at crime scenes, and the ammunition must be removed and the firearm rendered safe before it is packaged and transported. Since the item will likely be contained within a box or paper bag to protect it during transport, the dangerous nature of the item will likely be unknown by many of those who may handle it during lab submission. This is why extra care must be taken with items with the potential to cause harm during evidence collection, packaging, and transport.

1.3 Evidence Handling

Proper collection and packaging of evidence at the crime scene are also extremely important. The evidence can be easily lost or destroyed at this time, and the care and precision taken by the scene responders will dictate the success of the forensic testing. The results of any forensic testing are only as good as the evidence submitted. If evidence is carelessly collected and packaged, it will arrive at the lab in an altered state, which may prevent analysts from obtaining results from the items at all. A common phrase used in the field that describes this idea well is, "Garbage in, garbage out." Lab scientists are not able to perform miracles and restore destroyed evidence to usable form. Once that evidence is lost, there is no bringing it back and there is no telling what information could have been gained from it. If evidence collection and preservation techniques at the crime scene are substandard, the results will likely be substandard.

1.3.1 Evidence Recognition and Collection

Each evidence type encountered at a crime scene will require different considerations during the collection and packaging steps. For example, evidence in bombing and arson cases will be elementally analyzed for flammable components. Chemical accelerants used to start fires in arson cases may evaporate

Introduction 15

Figure 1.9 Metal paint cans are often used to store arson evidence as they prevent any volatile chemicals present from evaporating. (Source: Shutterstock.com)

quickly in the atmosphere. In order to trap these chemicals and prevent dissipation, arson evidence is often collected and stored in clean, metal paint containers which are airtight. The paint cans should be filled to at least 1/2 but no more than 2/3 fullness with debris from the scene. This space at the top of the can is necessary for subsequent analysis.

Drug evidence requires different considerations. Drugs are often collected and left in the packages in which they are recovered. This is because people who sell drugs have an interest in ensuring the substances remain in

Figure 1.10 A crew member aboard the Coast Guard Cutter Legare offloads seized cocaine at Coast Guard Base Miami Beach, Florida. (Source: U.S. Coast Guard photo by Petty Officer 3rd Class Mark Barney.)

Figure 1.11 White powder substance in a plastic bag. (Source: Shutterstock.com)

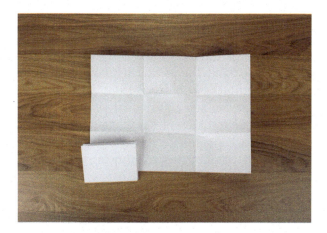

Figure 1.12 Demonstration of the proper way to fold glassine into a "druggist's fold." The evidence is placed in the middle of the paper. The paper is folded into thirds one way, and then the other way, creating an enclosed envelope around the evidence. The top flap is taped down and labeled. (Source: Shutterstock.com)

the package, so they take care to store them in something that will prevent loss from occurring. The powdered form of drugs such as cocaine will likely be stored in plastic bags or plastic bottles to guard against loss.

If a suspected drug is discovered at the crime scene and it is not contained within a package, it could also be stored in a druggist's fold and placed within a manila envelope for labeling and transport. The crime scene investigator would not want to collect powdered cocaine in a standard mailing envelope because the drug could be lost out the seams. Hairs and fibers should also be collected and placed in a druggist's fold before being stored in an envelope to prevent them from escaping through a seam.

Introduction

1.3.2 Evidence Packaging and Preservation

Biological evidence requires several collection and packaging considerations. When fresh body fluids are detected at a crime scene, items bearing biological material should be collected and packaged in cardboard boxes or paper bags. Items which will likely contain biological material, even if body fluids aren't indicated, should be packaged in paper and not plastic. This is because plastic traps moisture, which promotes growth of mold and bacteria. Mold and bacteria degrade biological material, including enzymes which are examined in forensic serology and DNA testing. If an item is collected from a crime scene, and biological testing is likely to be performed on the item, it should be packaged in a paper bag or a cardboard box.

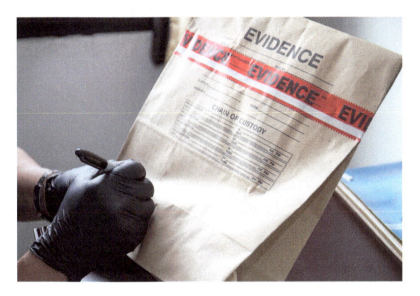

Figure 1.13 A paper evidence bag. (Source: Shutterstock.com)

Additionally, if items are wet from collection techniques, weather conditions, or body fluid saturation, they should be transported to the lab in paper/cardboard and then removed from the packaging once at the lab to allow the items to air dry. It is important to remove these items from the scene to prevent evidence loss and transport them to a controlled environment where the drying process can take place. Most labs have biological evidence lockers or fume hoods where evidence can be hung and dried.

Clean paper should be placed under the items to catch any trace evidence which may fall from the item during the drying process. Once the item is dry, it should be re-packaged and re-labeled with care (often including the original packaging and trace items) for storage until such time when it can be examined.

Figure 1.14 Evidence hanging to dry in a hood. (SecureDryTM evidence drying cabinet by MYSTAIRE; used courtesy, and with permission of, MYSTAIRE, www.mystaire.com)

All evidence items should be stored in a cool, dry place. If an item is packaged and stored properly, any evidence on it can exist safely for a number of years. DNA profiles have been successfully developed from items that were stored for 30 years or longer.

All evidence collected from the scene must be labeled and identified properly on the exterior packaging. This is the first step in maintaining an accurate chain of custody. The chain of custody is a formal record of all the places an item of evidence has been and all the people who have handled the item. A suitable chain of custody will show the dates and times when each scientist handled the item at each location. Every time an item of evidence is transferred from one area to another, or from one person to another, this must be listed on the chain of custody. Exterior packaging is an important link in the chain because the analyst who collected and packaged the item will place their initials on the package. The exterior packaging of any evidence item should bear the name

Introduction

or initials of the person who collected the evidence, the date the evidence was collected, and the item description and case number at a minimum. Different departments and labs may require additional information such as the time of collection and offense type. Ensuring that all evidence is sealed in properly labeled packages which are suitable for the item collected will protect against loss, contamination or deterioration of evidence during transport and storage.

1.3.3 Evidence Classes (Class/Individual)

In general, there are two main types of evidence: class and individual. Individual evidence is any item that has characteristics unique enough in nature to identify it to the exclusion of all other items. A fired projectile is an example of individual evidence if it has been imparted with markings from the interior of a gun barrel. A firearm can be test fired and a comparison can be made between the lands, grooves, and striations on the test-fired projectile and the lands, grooves, and striations from the evidence projectile. If the markings match, the projectiles were fired from the same gun. A bloodstain from a crime scene that yields a DNA profile is individual evidence because DNA profiles are unique from person to person. A footwear impression contains unique wear patterns and damage from wear which may be matched to the single shoe that made the impression.

Class evidence is that evidence that lacks unique characteristics and can never be traced to a source or matched to a person or item. Class evidence can be described as like or unlike, consistent or inconsistent with other evidence of its kind. Soil, fibers, glass, and hairs without roots are all examples of evidence that lacks unique characteristics. Class evidence is often mass produced, like panes of glass or textiles. Footwear impressions from new shoes and ballistic evidence with insufficient minutiae are also considered class evidence because they can never be traced back to the specific shoe that made the impression or the gun that fired the projectile.

1.4 Crime Laboratory Operations—Overview

There are hundreds of public crime laboratories in the United States, each serving communities with different needs and challenges. Some labs serve large, rural geographical areas and some serve small, condensed, highly populous cities. There are four federal forensic labs that serve the United States. The Federal Bureau of Investigation (FBI) lab is widely regarded as the premier forensic laboratory in the country. The FBI lab processes forensic evidence dealing with federal criminal law violations and also assists state laboratories in the processing of evidence which they are not trained or equipped to process. The other federal crime labs include the Drug Enforcement Agency lab, the Alcohol, Tobacco, and Firearms labs, and the United States Postal Service lab.

Each state has at least one public crime lab and some states have a system of public crime labs that cover a large geographical area, like the states of California and Florida. Most labs offer testing in forensic biology, forensic chemistry, trace evidence analysis, toxicology, latent fingerprints and friction ridge comparison, questioned documents, fire debris and bombs/arson examination, firearms and toolmark comparison, and digital evidence.

1.4.1 Forensic Biology

Forensic biology is comprised of two primary disciplines: serology and DNA analysis. Serology is the study of human body fluids. At some labs, the serological examination takes place in a separate unit or by a separate analyst. At other labs, the serological examination and DNA testing are performed by the same analyst. The goal of this testing is to identify whether biological body fluids are present. A complete biological examination includes a visual examination with the naked eye, examination under an alternate light source, and chemical testing to determine the presence and location of biological material. If biological material has been detected or is suspected to be present, the serologist then prepares samples to send forward for DNA testing. There are a number of chemical analyses that can be performed to characterize body fluids, but the tests fall into two major groups: presumptive and confirmatory tests. Some combination of presumptive and confirmatory tests take place in many disciplines of forensic science on both the biology and chemistry sides. We will discuss how the presumptive and confirmatory tests in forensic biology work in Chapter 4. Common evidence types in forensic biology are blood, semen, saliva, sweat, and skin cells.

1.4.2 Controlled Substances

Controlled substances analysis is the study of illegal drugs. Drugs are substances that can be defined as any natural or synthetic substance that is used to produce a physiological effect. Certain drugs are illegal because there are laws at the federal and state level that prohibit their use or distribution. Controlled substances are often examined with a number of different tests, starting with the screening test and ending with confirmatory tests. Screening tests are often color change or microcrystalline tests. One or more of these screening tests can be applied to narrow the pool of possibilities before moving to confirmation.

Confirmatory tests are used to identify the drug and determine its concentration if necessary. Most often infrared spectroscopy and gas chromatography/mass spectrometry are used to confirm the identity of a drug or compound.

Introduction

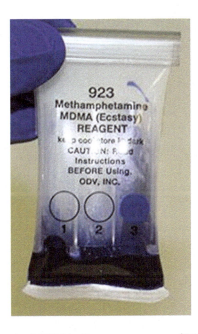

Figure 1.15 Color test for MDMA. (Image courtesy of NFSTC@FIU.)

There are a number of drug classifications: one such classification is narcotics. Though illicit drugs are often referred to synonymously as "narcotics," clinical narcotics relieve pain by depressing the central nervous system. Narcotics can affect blood pressure, pulse rate, and breathing rate. The regular use of narcotics will result in physical dependence by the user. The most commonly abused narcotics are opiates, which are derived from the poppy.

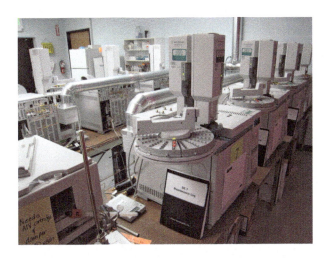

Figure 1.16 Gas chromatography laboratory. (Photo retrieved from Wikimedia, used per Creative Commons 2.0, user Hey Paul from Sacramento, CA, USA.)

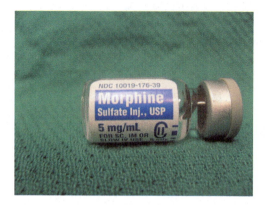

Figure 1.17 Vial of morphine. (Photo retrieved from Wikimedia, used per Creative Commons 3.0, user Vaprotan.)

Morphine, and drugs derived from morphine, are the most commonly encountered opiates, though pharmaceutically produced opiates have been featured prominently in the rise of the current opiate abuse epidemic. Morphine has medical application for pain management in a clinical setting. Heroin, which is derived from morphine, is an opiate that abusers dissolve in water and inject intravenously or subcutaneously. Heroin produces a euphoric high which lasts for three to four hours accompanied by drowsiness. Codeine, oxycodone and methadone are other commonly encountered opiates.

Figure 1.18 Preparing heroin for injection by boiling. (Photo retrieved from Wikimedia, used per Creative Commons 3.0, user Hendrike.)

Hallucinogens are another classification of drugs. These substances interrupt normal thought, perception, and mood. The most frequently encountered drug in this class is marijuana, which is derived from the cannabis plant. A substance present in the plant, tetrahydrocannabinol (THC), is responsible for the hallucinogenic properties of marijuana. Few studies have been performed on the long-term effects of marijuana use or the development of

Introduction 23

physical or psychological dependence on the drug. Other commonly encountered hallucinogens are LSD, PCP, MDMA (Ecstasy, Molly), and psilocin and psilocybin in "magic" mushrooms.

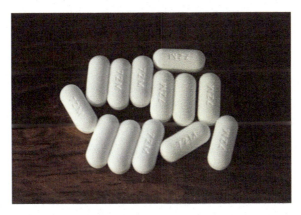

Figure 1.19 Pills containing codeine. (Photo retrieved from Wikimedia, used per Creative Commons 3.0, user ZngZng.)

Figure 1.20 Cannabis sativa. (Photo retrieved from Wikimedia, used per Creative Commons 3.0, user Bogdan.)

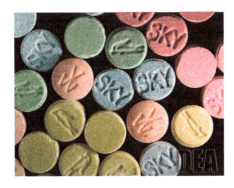

Figure 1.21 Ecstasy tablets. (Source: Drug Enforcement Agency (DEA).)

Figure 1.22 Hallucinogenic mushrooms.

Depressants are another classification of drugs. Narcotics can be depressants, but not all depressants have medical application for pain management. These drugs depress the function of the central nervous system. The most commonly encountered depressant in forensic science is alcohol. Alcohol, a legal drug, is consumed most often by mouth. It enters the bloodstream, travels to the brain and depresses the body's normal thought processes or muscle control.

Aside from alcohol (also called ethyl alcohol or ethanol), other commonly encountered depressants are barbiturates, tranquilizers, and even household products such as glues and aerosols which can be inhaled to produce the effect.

Figure 1.23 Alcohol is the most widely abused drug in the Western hemisphere. (Photo retrieved from Wikimedia, used per Creative Commons 4.0 International, user Tiia Monto.)

Introduction

Stimulants are a class of drug which stands in opposition to the depressants, as they stimulate the activity of the central nervous system. Drugs of this type increase alertness or activity in the user. The most commonly encountered stimulants are cocaine and methamphetamine. Some stimulants, like methamphetamine, simultaneously produce intense feelings of pleasure along with increased alertness or activity in the user. Cocaine, which is derived from the coca plant, suppresses hunger, fatigue, and boredom, and increases alertness or vigor. A potent form of cocaine, which is mixed with baking soda and water and heated, is called "crack." Crack is often smoked from a glass pipe and produces similar effects as the other stimulants.

Figure 1.24 Cocaine in powder form. (Photo retrieved from Wikimedia, used per Creative Commons 3.0, user Davidfernandocoronel.)

Club drugs and date rape drugs are another class of illegal drugs. These are most often used at electronic music festivals, clubs, and raves. These drugs are classified as hallucinogens or depressants, and are often used to facilitate sexual assault. Examples of club drugs are Ecstasy or Molly (MDMA), GHB (gamma hydroxybutyrate), and Roofies (Rohypnol).

The aforementioned drugs have been described and classified by the federal government in the Federal Controlled Substances Act, established in 1970. The Act places drugs into five categories, or schedules, based on the drug's potential for abuse, potential for physical and psychological dependence, and medical value.

The drugs in Schedule I have a high potential for abuse and no currently recognized medical value. Examples of drugs in Schedule I are heroin, marijuana, methaqualone, and LSD. The drugs in Schedule II have a high potential for abuse and limited medical value, which is tightly restricted. They include cocaine, PCP, amphetamines, and barbiturates. Schedule III drugs have a reduced potential for abuse and a currently recognized medical application, such as codeine (over a certain amount), anabolic steroids,

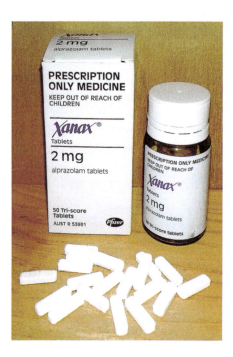

Figure 1.25 Xanax, a popular barbiturate.

and all barbiturates that are not addressed in Schedule II. Schedule IV drugs have a low potential for abuse with a currently recognized medical application and these include some tranquilizers such as Valium, Ambien, and Ativan. Schedule V drugs show low potential for abuse with a recognized medical application, such as cough syrups containing codeine under a certain amount and the cough suppressant guaifenesin.

Figure 1.26 Methamphetamine in crystal form. (Photo retrieved from Wikimedia, used per Creative Commons 4.0 International, user Radspunk.)

Introduction 27

Figure 1.27 Molly, a club drug that can also be used in drug-facilitated sexual assault. (Source: Shutterstock.com)

Figure 1.28 GHB and Roofies are known date rape drugs. (Source: Shutterstock.com)

28   Guide to the ABC Biology Exam

Table 1.1   How Drugs Are Classified in the United States

Schedule	Description	Examples
Schedule I	This schedule consists of drugs with no currently accepted medical use and a high potential for abuse. They are the most dangerous drugs of all the schedules, and have potentially severe psychological or physical dependence.	• Heroin • Lysergic acid diethylamide (LSD) • Marijuana • MDMA (Ecstasy) • Methaqualone • Peyote
Schedule II	This schedule includes drugs with a high potential for abuse, with use potentially leading to severe psychological or physical dependence. These are considered dangerous drugs.	• Cocaine • Methamphetamine • Methadone • Hydromorphone • Meperidine • Oxycodone • Fentanyl • Dexedrine • Adderall • Ritalin • Combination products with less than 15mg of hydrocodone per dosage unit
Schedule III	This schedule contains drugs with a moderate to low potential for physical and psychological dependence. These drugs have a lower abuse potential than the drugs in Schedules I and II, but a higher abuse potential than those in Schedules IV and V.	• Ketamine • Anabolic Steroids • Testosterone • Products containing less than 90mg of codeine per dosage unit
Schedule IV	This schedule includes drugs with a low potential for abuse and low risk of dependence, but not as low a risk as drugs in Schedule V.	• Xanax • Soma • Darvon • Darvocet • Valium • Ativan • Talwin • Ambien • Tramadol
Schedule V	This schedule contains drugs with lower potential for abuse than drugs in Schedule IV and consists of preparations containing limited quantities of certain narcotics. Schedule V drugs are generally used for antidiarrheal, antitussive, and analgesic purposes.	• Cough preparations with less than 200mg of codeine per 100mL • Lomotil • Motofen • Lyrica • Parepectolin

Introduction

1.4.3 Toxicology

Because of the prevalence of drugs in modern society, forensic toxicology has become a prominent discipline in the field of forensic science. Toxicologists detect and identify drugs and poisons in the body. They test body fluids, such as blood, tissues, and organs to test for the presence of drugs or their metabolites. Toxicological tests are required in crime laboratories and medical examiners' offices, but they are also necessary in hospital laboratories and health facilities responsible for monitoring the intake of drugs and other toxic substances.

The role of the forensic toxicologist is to perform toxicological tests as they pertain to violations of criminal law. Sometimes this testing is performed in a crime laboratory independent of the medical examiner. With smaller systems, this responsibility may reside with one or the other or may be handled by a governmental health department laboratory.

One of the highest volume tests performed by the forensic toxicologist is blood alcohol testing. Ethyl alcohol is the most heavily abused drug in the Western hemisphere. Blood alcohol testing and controlled substances testing remain two of the highest volume sections of the modern crime lab. Other common toxicological examinations look for the presence of controlled substances or poisons in the body.

1.4.4 Trace Analysis

Trace Analysis is the examination of hairs, fibers, paints, and particles that are transferred during the commission of a crime. Trace analysis most often involves examination of trace materials under a microscope to determine if they are similar or dissimilar. Materials recovered from a crime scene can be examined and compared to reference collections. This is done in order to identify the material and also to determine if it is similar or dissimilar to other trace materials in and around the crime scene. Common types of trace evidence include broken glass, soil, dirt, human and animal hair, natural and synthetic fibers, and paints and polymers.

1.4.5 Latent Fingerprints

Before the advent of forensic DNA analysis, there was only one type of evidence that could identify a suspected perpetrator by name and conclusively place him or her at the scene of a crime. Fingerprints join DNA as one of the few types of evidence that is considered individual, meaning evidence unique enough to identify a source to the exclusion of all others. There are three foundational beliefs of friction ridge analysis: fingerprints are unique, they can be classified, and they remain unchanged over a person's lifetime. These principles allow fingerprints to be analyzed and matched back to the person who left them behind, if there is enough detail.

There are two main types of fingerprints: patent and latent. Patent fingerprints can be seen with the naked eye and require no additional methods to visualize them. These can be plastic prints, which are prints impressed in a pliable surface such as wax, soap, or dust.

Figure 1.29 Patent fingerprint left in blood. (Source: Shutterstock.com)

Patent prints can also be visible when someone touches an item that bears blood, grease, dye, or paint and then touches another item, leaving behind the details of their fingerprint like a stamp after it has contact with an inkpad. The opposite of patent fingerprints are latent fingerprints. The fine details of a latent fingerprint can't initially be visualized fully with the naked eye. In the case of latent fingerprints, some additional treatment is required in order to fully visualize the details that are ultimately examined in friction ridge analysis and comparison.

Latent fingerprints are visualized in a number of different ways at a crime scene and in the forensic lab. The method of detection for latent fingerprints is largely based on the qualities of the surface from which the print is being lifted. For non-porous surfaces, the most common method for visualizing latent fingerprints is by dusting with powder. The powder is applied to the fingerprint using a brush, and adheres to the sweat and oils in the fingerprint, making it visible. Powders can also vary based on the characteristics of the surface from which the print is being collected. The color of the powder used to visualize the latent print should contrast with the color of the surface from which the print is being collected. For example, if fingerprints are being collected from a white surface, a dark powder should be used to treat them.

Figure 1.30 Latent fingerprint visible after enhancement with black powder. (Source: Shutterstock.com)

If they are suspected on a black surface, a white or silver powder should be employed.

Another commonly used method for detection of latent fingerprints on non-porous surfaces is cyanoacrylate (also referred to as superglue) fuming. On porous surfaces, like paper, one of the most common ways to develop latent fingerprints is to use ninhydrin spray. The chemicals in the ninhydrin spray react with the amino acids and salts in the fingerprint residue to produce a purple color. This method is most often used when developing latent fingerprints that were deposited on paper items, such as envelopes, bank notes, and checks. Other methods that can be attempted with porous surfaces are iodine fuming and development using silver nitrate.

There are other, less commonly used processes for developing latent fingerprints. For example, Amido Black, Leuco Crystal Violet, Leuco Malachite Green, and Coomassie Blue are chemical developers that are used to develop fingerprints in blood.

There are three main types of fingerprint patterns: loops, whorls, and arches. Within the ridges and valleys on a person's fingers are many other

Figure 1.31 Loop pattern. (Source: NIST.)

Figure 1.32 Whorl pattern. (Source: NIST.)

types of characteristics, called minutiae, which make a person's print unique. The placement of ridge endings and bifurcations in a crime scene fingerprint can be examined and compared to the fingerprint characteristics of a known individual to determine if they match. Fingerprints from crime scene evidence can be compared with the fingerprint from a suspected person or they can be run through a database to determine if a match exists.

Introduction

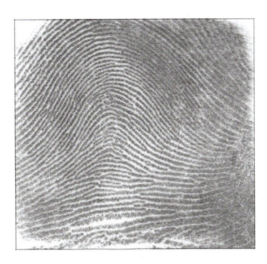

Figure 1.33 Arch pattern. (Source: NIST.)

1.4.6 Questioned Documents

Questioned document examination is a part of the lab that deals with the examination of a number of different evidentiary samples that may be important to a criminal investigation. Examiners perform forgery analysis, which can include alternate light source and chemical analysis of inks and dyes to determine similarity or continuity. They also perform handwriting analysis and comparison between exemplar writing samples and questioned documents to ascertain whether they had the same author. Questioned document examiners analyze documents damaged by water and fire to visualize writing that existed on the item before the damage occurred. They process items, such as notepads, which may exhibit indented writing samples that could lend clues as to what was written on used and removed pages of the pad. Questioned document examiners also examine printing artifacts to compare with printers, copy machines, and typewriters to determine similarities and differences as they relate to the investigation.

1.4.7 Fire Debris

Fire debris analysis involves detection of ignitable fluid residues at suspected arson scenes. Charred items from the fire scene are transported back to the lab in airtight containers, most often new, clean paint cans or glass jars. Investigators leave a space in the evidence collection container, one-half to one-third of the volume of the can or jar, where gases can escape from the evidence for analysis. This airspace left in the cans is called "headspace" and contains any volatile substances that have evaporated off the evidence. This "headspace" can then be extracted or concentrated onto a charcoal tab for examination through elemental analysis using liquid or gas chromatography and mass spectrometry. Fire investigators can examine items from fire scenes to determine points of origin. They can look at items located in different areas of a building to determine which areas burned hottest and longest, as well as how the fire spread through the structure. Common fire debris evidence types are wood, carpets and cushions in and around the suspected accelerant area, containers that may have contained an accelerant, and ignition devices, such as matchbooks.

Figure 1.34 Matches are a commonly used ignition device in arson cases. (Photo retrieved from Wikimedia, used per Creative Commons 3.0, user StoatBringer.)

1.4.8 Firearms and Toolmarks

Firearms analysis is the collective term used to describe the examination and comparison of markings on projectiles and cartridge casings with the goal of identifying which weapon made the markings, and therefore fired the bullet. Common evidence in firearms examination labs are the firearms themselves, whole bullet cartridges, fired projectiles, and fired casings. The cartridge, projectile, and casing will bear markings made by the barrel of the gun and the firing mechanisms inside the gun when the projectile was fired. For example, the rifling marks that are impressed on a projectile when it is fired through the barrel of the gun include the lands, grooves, and striations imparted to the barrel during the manufacturing process of the firearm. The casings are examined for marks made by the firing pin, ejector, and extractor of the firearm. Even a cartridge that is unfired may bear some markings made on it by simply being loaded into the chamber of the firearm. Just like for DNA and fingerprints, there are databases maintained for known and questioned markings and these can be searched for similarity between spent projectiles and casings from other crime scenes and suspect weapons.

In some laboratories, ballistic examination, or the study of a projectile in motion, would fall under the discipline of firearms analysis. This area of

Introduction

analysis deals with bullet trajectory and crime scene reconstruction. Gunshot residue detection also falls under the umbrella of firearms analysis, and this involves not only the detection of the burned and unburned powder particles, but also characterization and test firing for shot distance determination. Moreover, firearms examination includes the analysis of the firearm to raise obliterated serial numbers.

Figure 1.35 Ceramic chest plates demonstrating the appearance of wound patterns from a Smith & Wesson .38 revolver shot at various distances. (Source: NIH, Office of the Chief Medical Examiner, Baltimore, Maryland. From exhibition "Visible Proofs: Forensic Views of the Body" U.S. National Library of Medicine, 8600 Rockville Pike, Bethesda, MD 20894.)

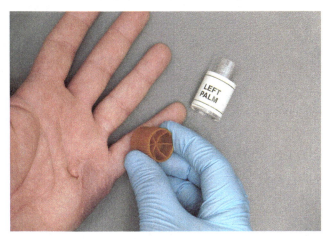

Figure 1.36 Gunshot residue collection stubs used to collect particles from a shooter's hands. (Source: Shutterstock.com)

Figure 1.37 (Left) Before and (Right) after images of a firearm serial number restoration. (Used with permission of Iowa Department of Public Safety, Division of Criminal Investigation.)

Collection and preservation of firearms evidence can be complicated and is subject to some unique circumstances. Firearms must first be rendered safe by a qualified firearms expert before they are packaged. They are usually secured within a cardboard box with zip ties or fasteners to hold them in place. If a firearm is recovered from water, it should be stored and transported to the lab in the same water from which it was recovered to prevent rust from forming. Rust can significantly affect the markings made by the weapon during test fire, so it's of paramount importance that the firearm be collected and preserved in such a way that will prevent rust from forming. Also, nothing should ever be placed in the barrel of a firearm for the purposes of lifting and collecting it.

Tool marks are any gouge, abrasion, or cut caused by a tool coming into contact with another object. Typically, the tool must be applied to the surface with sufficient force to impart a mark on the medium, and the medium must be soft enough to be impressed with the tool mark. Though most tool marks will likely always bear class characteristics of the tool that made them, including the type and size, sometimes there are individual characteristics unique to that particular tool or weapon. These individual characteristics are defects or damage from routine use or wear that leave impressions which are unlike any other. Tool marks are often collected by casting the impression with silicone gel or paste, such as AccuTrans.

1.4.9 Digital Evidence

Digital evidence analysis involves examining computers, cell phones, images, audio and digital recordings, and files. As society relies more and more on computers for all aspects of daily life, so too do we rely on forensic scientists to collect and analyze digital information to help solve crime. Digital analysis includes the examination of computer hardware and software, files and

Introduction

emails, internet searches, and data storage. Digital evidence can be found on desktop and laptop computers, external hard drives and flash drives, digital cameras, and mobile phones.

Notes

1. American Board of Criminalistics (August 2017). *Molecular biology exam study guide*. Retrieved from http://www.criminalistics.com/study-guides.html.
2. Saferstein, R. (2018) *Criminalistics: an introduction to forensic science* (12th ed.). Upper Saddle River, N.J.: Pearson.
3. Ibid.
4. Ibid.
5. Ibid.
6. Ibid.
7. Ibid.
8. Ibid.
9. Ibid.
10. McCrone Research Institute/Chicago (n.d.). Retrieved November 18, 2019, from https://www.mccroneinstitute.org/.
11. Saferstein, R. (2018) *Criminalistics: an introduction to forensic science* (12th ed.). Upper Saddle River, N.J.: Pearson.

Quality Assurance and Quality Control

2

2.1 Quality Assurance and Quality Control

Quality assurance is a program put in place at laboratories to help ensure the production of accurate and reliable test results. Quality control procedures are the multitude of checks and balances that maintain the program of quality assurance. Quality assurance covers the accreditation of the lab, the proficiency and competency of the analysts, the calibration of the instrumentation, the validations of all methods, and the purity of the consumable equipment and reagents used during the testing. Quality control procedures are the means by which the quality of the laboratory results is assured.

2.1.1 Accreditation

Accreditation is a vital part of any quality assurance program. Accreditation is the formal recognition of a laboratory, by a private accrediting agency, acknowledging compliance with recognized industry standards and best practices. This is accomplished by way of audits of the laboratory, both internal and external, to establish initial and ongoing compliance with these standards. The accrediting agency initially visits the laboratory for a period of days and auditors review the protocols and procedures, management systems, and quality systems in place to ensure that they are in line with established industry standards. Audits are performed periodically each year, rotating through some combination of internal audits by the laboratory personnel and external audits by the accrediting body.

During the audits, the accrediting body will review laboratory protocols and procedures, quality control documentation, analyst proficiency documentation, laboratory testing notes, and chain of custody documents. Auditors will review samples of technical data and reports to ensure analysts are following the protocols in place. They examine the educational qualifications, training, and proficiency documentation of each analyst to ensure they are qualified to perform the testing, and they also review sample tracking, inventory, and storage documents. They interview analysts and observe laboratory activities to see that all protocols are being followed. Everything that takes place in the laboratory, from the receipt of the reagents used in the testing process, to their assessment of suitability for use in testing, to the storage conditions and equipment maintenance, is documented in some form for review. The documentation of the laboratory activity for both testing and quality assurance and the preservation of that documentation for audit review is known as the creation of an audit trail. An audit trail refers to the documents that an auditor or other qualified analyst could review to ascertain what tests were performed, when they were performed, and what the results of those tests showed.

2.1.2 Accrediting Bodies

Historically, there were two major accrediting bodies for forensic laboratories in the United States: The American Society of Crime Laboratory Directors/Laboratory Accreditation Board (ASCLD/LAB) and Forensic Quality Services (FQS). ASCLD/LAB was established in 1984 and was the first body to accredit forensic labs in the United States. Some states have legislation providing that public and/or private labs must be accredited by ASCLD/LAB (e.g. New York State). In April 2016, ASCLD/LAB was acquired by the ANSI National Accreditation Board (ANAB). The other major forensic accrediting body was Forensic Quality Services (FQS). Forensic Quality Services was established in 2004 and was acquired by ANAB in November of 2011. Though FQS wasn't the first accrediting body, it was the first body to incorporate international standards into their accreditation requirements. Standards established by the International Organization for Standardization (ISO) were first implemented into FQS audits, which required laboratories to adhere to these international standards as well as the established federal and agency specific best practices. The international standards have gone on to be the gold standard in forensic accreditation and now all forensic laboratories have some combination of domestic and international elements.

As it stands today, the major accrediting agency in forensic science in the United States is ANAB. With both ASCLD/LAB and FQS under its umbrella, ANAB now accredits the vast majority of forensic labs in the US (www.anab.org).

Laboratories voluntarily seek out these private agencies to evaluate the protocols and procedures in place at the lab and determine whether the lab is in line with the accrediting body's standards. The accrediting bodies incorporate federal and international standards that apply to forensic laboratories, and also, at times, standards specific to the agency, which have been recognized by the auditors and board members as being important for maintaining the quality and integrity of the analysis in a forensic setting. In forensic biology and DNA analysis, some of these standards include the FBI Quality Assurance Standards (QAS), which were established in 1998 and later amended in 2007, 2008, and 2011. These standards were derived from the DNA Advisory Board (DAB), which was established by the director of the FBI under the DNA Identification Act of 1994. The Board first convened

Quality Assurance and Quality Control

in 1995 and was disbanded in 2000, and in that time they made recommendations to the FBI regarding proficiency testing for DNA labs, DNA profile guidelines for entry into the CODIS database, certification and educational requirements for DNA analysts, and statistical analysis of forensic DNA profiles. Since the DAB was disbanded, the Scientific Working Group on DNA Analysis Methods (SWGDAM) now provides recommendations to the FBI director regarding the QAS in order to maintain up-to-date standards as the practice evolves. Any forensic biology lab that enters DNA profiles into CODIS must adhere to the FBI's Quality Assurance Standards, and be audited against these standards.

Forensic accrediting bodies have also now incorporated some international standards, such as the ISO 17025 standard. One of the applications of the ISO 17025 standard is for analytical testing laboratories. ISO 17025 focuses on accuracy, precision, metrological traceability, measurement uncertainty, proficiency, and method validation in forensic testing laboratories.

Figure 2.1 The International Organization for Standardization is the primary international forensic standards body. (Used courtesy, and with permission of, ISO.)

2.1.3 Personnel Certification

Where *accreditation* applies to forensic laboratories, *certification* applies to individual forensic analysts. "Certification is a voluntary process of peer review by which a practitioner is recognized for attaining the professional qualifications necessary to practice in one or more disciplines of criminalistics" (www.criminalistics.com). The goals of forensic certification are "to set and measure professional levels of knowledge, skill and abilities, to guide professionals in the attainment of professional levels of competence, to provide a means of evaluating the competence of practitioners, and to provide a formal process for the recognition of practitioners who meet the professional level of competence" (www.criminalistics.com). Certification is achieved through a series of qualifying evaluations, including a review of the analyst's

Figure 2.2 The American Board of Criminalistics offers forensic analyst certification in six subdisciplines: comprehensive criminalistics, drug analysis, molecular biology, fire debris analysis, hairs and fibers, and paints and polymers. (Used courtesy, and with permission of, ABC.)

education, training, and experience, a written examination in the discipline of practice, and ongoing involvement and continuing education in the discipline of practice as well as in forensic science in general.

One of the major certification bodies that exist in forensic science is the American Board of Criminalistics (ABC), which offers certification in general criminalistics as well as in several specialty areas of practice, which include drug analysis, molecular biology, fire debris analysis, hairs and fibers, and paints and polymers.

The International Association of Arson Investigators (IAAI) offers several levels of certification for arson investigators.

The IAAI, which has over 8,000 members worldwide, offers certified fire investigator, fire investigation technician, evidence collection technician, and certified instructor certifications.

The International Association for Identification (IAI) is another agency that offers certification for forensic practitioners.

The process of obtaining certification through the IAI is similar to that of the ABC or IAAI, although some of the certification disciplines also include a practical component. Some of the areas of practice that the IAI offers certification are latent print examination, crime scene photography, crime scene reconstruction, and bloodstain pattern analysis.

Figure 2.3 The International Association for Identification is a main certifying body for forensic field disciplines. (Used courtesy of IAI.)

Quality Assurance and Quality Control 43

Figure 2.4 The American Board of Forensic Toxicology offers forensic certification to toxicological examiners. (Used courtesy, and with permission of, ABFT.)

The American Board of Forensic Toxicology (ABFT) was founded in 1975 with the objective of "establishing, enhancing, and revising as necessary the standards of qualification for those who practice forensic toxicology and to certify as qualified scientists those voluntary applicants who comply with the requirements of the Board" (www.ABFT.org). As of this writing, there were 471 certificants of the ABFT across four different certifications.

The Association of Firearm and Tool Mark Examiners (AFTE) is another prominent body that certifies forensic analysts in the area of firearms analysis and tool mark analysis. The stated purpose of the AFTE certification program is to "demonstrate to interested parties that successful applicants have met a standard of excellence in knowledge and skill for a qualified examiner as defined by AFTE and also to promote professionalism among firearm and toolmark examiners by establishing certification as a level of accomplishment" (www.afte.org).

A few other certification bodies that are noteworthy in forensic science include the American Board of Forensic Document Examiners (ABFDE),

Figure 2.5 The Association of Firearm and Tool Mark Examiners offers certification that consists of written and practical examinations in firearms analysis, tool mark analysis, and gunshot residue examination. (Courtesy of AFTE.)

Figure 2.6 This table lists all of the known forensic certifying bodies and was compiled by the National Commission on Forensic Science as part of their views document on the certification of forensic science practitioners, published in 2016. (Source: NIST; National Commission on Forensic Science.)

ILAC G19 Categories of Testing	Discipline/ Subdiscipline	Certification Organization	Contact Info	Accrediting Organization	Approximate No. of Diplomats or Certified Individuals	Application Fee	Examination Fee	Annual Recertification Fee or Dues
Controlled Substances	Drug Analysis	American Board of Criminalistics	www.criminalistics.com	FSAB	232	$50	$250	$50
Comprehensive Criminalistics Examination (General Criminalistics)	Criminalistics Subjects	American Board of Criminalistics	www.criminalistics.com	FSAB	111 (718)	$50	$250	$50
Hairs, Blood, Body Fluids and Tissues	Molecular Biology	American Board of Criminalistics	www.criminalistics.com	FSAB	364	$50	$250	$50
Trace Evidence	Fire Debris	American Board of Criminalistics	www.criminalistics.com	FSAB	47	$50	$250	$50
Trace Evidence	Hairs and Fibers	American Board of Criminalistics	www.criminalistics.com	FSAB	27	$50	$250	$50
Trace Evidence	Paint and Polymers	American Board of Criminalistics	www.criminalistics.com	FSAB	22	$50	$250	$50
Entomology	Forensic Entomology	American Board of Forensic Entomology	www.forensicentomologist.org		16	$50	$50	$50
Handwriting and Document Examination	Forensic Document Examination	American Board of Forensic Document Examiners	www.abfde.org	FSAB	106	$250	N/A	$250 Annual Dues
Handwriting and Document Examination	Forensic Document Examination	Board of Forensic Document Examination	www.bfde.org	FSAB	14	$100	$500	$50
Fingerprints	Latent Fingerprints	International Association for Identification	www.theiai.org	FSAB	1041	$200/IAI Members; $300/Non-Members	N/A	Recert every 5 yrs: $200/IAI Members; $300/Non-Members

Quality Assurance and Quality Control

ILAC G19 Categories of Testing	Discipline/ Subdiscipline	Certification Organization	Contact Info	Accrediting Organization	Approximate No. of Diplomats or Certified Individuals	Application Fee	Examination Fee	Annual Recertification Fee or Dues
Fingerprints	Ten-Print Fingerprints	International Association for Identification	www.theiai.org	FSAB	117	$200/IAI Members; $300/Non-Members	N/A	Recert every 5 yrs: $200/IAI Members; $300/Non-Members
Scene Investigation	Blood Stain Pattern	International Association for Identification	www.theiai.org	FSAB	39	$200/IAI Members; $300/Non-Members	N/A	Recert every 5 yrs: $200/IAI Members; $300/Non-Members
Scene Investigation	Crime Scene-Four Levels	International Association for Identification	www.theiai.org	FSAB	1,625: Outside FSSPs	$200/IAI Members; $300/Non-Members	N/A	Recert every 5 yrs: $200/IAI Members; $300/Non-Members
Scene Investigation	Forensic Artist	International Association for Identification	www.theiai.org	FSAB	34	$200/IAI Members; $300/Non-Members	N/A	Recert every 5 yrs: $200/IAI Members; $300/Non-Members
Scene Investigation	Forensic Photography	International Association for Identification	www.theiai.org	FSAB	57	$200/IAI Members; $300/Non-Members	N/A	Recert every 5 yrs: $200/IAI Members; $300/Non-Members
Marks and Impressions	Footwear	International Association of Identification	www.theiai.org	FSAB	105	$200/IAI Members; $300/Non-Members	$300	$200/IAI Members; $300/Non-Members

Figure 2.6 (Continued)

ILAC G19 Categories of Testing	Discipline/ Subdiscipline	Certification Organization	Contact Info	Accrediting Organization	Approximate No. of Diplomats or Certified Individuals	Application Fee	Examination Fee	Annual Recertification Fee or Dues
Audio, Video and Computer Analysis	Digital Evidence/ Video—Forensic Video Certification	International Association of Identification	www.theiai.org		23	$200/IAI Members; $300/Non-Members	$300	$200/IAI Members; $300/Non-Members
Marks and Impressions	Footwear/ Fingerprints	Canadian Identification Society	www.cis-sci.ca		N/A for U.S.A.	$150	N/A	$150
Audio, Video and Computer Analysis	Digital Evidence/ Video—Certified Forensic Video Analyst	Law Enforcement and Emergency Services Video Association	www.leva.org		54	N/A	N/A	$55/year
Audio, Video and Computer Analysis	Digital Evidence/ Video—Certified Forensic Video Technician	Law Enforcement and Emergency Services Video Association	www.leva.org		267	N/A	N/A	$55/year
	Evidence Handling	International Association for Property and Evidence	www.IAPE.org		1,400+; Outside FSSP	$150	N/A	$100
Firearms and ballistics	Firearms	Association of Firearm and Tool Mark Examiners	www.afte.org		116	N/A	$250	$25 every 5 years
Marks and Impressions	Tool Marks	Association of Firearm and Tool Mark Examiners	www.afte.org		46	N/A	$250	$25
Firearm Distance Determination	Gunshot Residue	Association of Firearm and Tool Mark Examiners	www.afte.org		39	N/A	$250	$25

Figure 2.6 (Continued)

Quality Assurance and Quality Control

ILAC G19 Categories of Testing	Discipline/ Subdiscipline	Certification Organization	Contact Info	Accrediting Organization	Approximate No. of Diplomats or Certified Individuals	Application Fee	Examination Fee	Annual Recertification Fee or Dues
Audio, Video and Computer Analysis	Digital Evidence/ Computer Forensics—Digital Forensics Certified Practitioner and DFCA	Digital Forensics Certification Board	www.dfcb.org		178	$250	$100	N/A
Audio, Video and Computer Analysis	Digital Evidence/ Computer Forensics—Certified Computer Examiner	International Society of Forensic Computer Examiners	www.isfce.org		805	$395	N/A	$75
Audio, Video and Computer Analysis	Digital Evidence/ Computer Forensics—Certified Digital Forensic Examiner, Certified Digital Media Collector, Certified Computer Crime Investigator	DOD Cyber Crime Center	www.dc3.mil		Training source	N/A	N/A	N/A
Audio, Video and Computer Analysis	Digital Evidence/ Computer Forensics—Certified Forensic Computer Examiner	International Association of Computer Investigative Specialists	www.iacis.com	FSAB	1,963	N/A	w/ training ($2,795) wo/ training ($750)	$50
Audio, Video and Computer Analysis	Digital Evidence/ Computer Forensics—Certified Advanced Windows Forensic Examiner	International Association of Computer Investigative Specialists	www.iacis.com		26	N/A	w/ training ($1,495) wo/ training ($750)	$50

Figure 2.6 (Continued)

ILAC G19 Categories of Testing	Discipline/ Subdiscipline	Certification Organization	Contact Info	Accrediting Organization	Approximate No. of Diplomats or Certified Individuals	Application Fee	Examination Fee	Annual Recertification Fee or Dues
Audio, Video and Computer Analysis	Digital Evidence/ Mobile Devices— Certified Mobile Device Examiner	International Association of Computer Investigative Specialists	www.iacis.com			N/A	w/ training ($1,495) wo/ training ($750)	$50
Audio, Video and Computer Analysis	Digital Evidence/ Mobile Devices— Advanced Smartphone Forensics	Global Information Assurance Certification	www.giac.org	ANSI	GOAC number certified is not available, but 80,079 certifications granted	N/A	w/training ($1,149) wo/training ($659)	$399
Audio, Video and Computer Analysis	Digital Evidence/ Computer Forensics— Certified Forensic Analyst, Certified Forensic Examiner, Reverse Engineering Malware, many others	Global Information Assurance Certification	www.giac.org	ANSI	GOAC number certified is not available, but 80,079 certifications granted	N/A	w/training ($1,149) wo/training ($659)	$399
	Forensic Engineering	International Board of Forensic Engineering Sciences	www.iifes.org	FSAB	16; Outside of FSSP	$300	N/A	$50
	Forensic Engineering	National Academy of Forensic Engineers	www.nafe.org	Council of Engineering and Scientific Specialty	Outside of FSSP (3 :3 Board Certified)	$125	N/A	$200–$300
	Civil Engineering	American Society of Civil Engineers	www.asce.org	ANSI	Outside of FSSP	N/A	N/A	N/A

Figure 2.6 (Continued)

Quality Assurance and Quality Control

ILAC G19 Categories of Testing	Discipline/ Subdiscipline	Certification Organization	Contact Info	Accrediting Organization	Approximate No. of Diplomats or Certified Individuals	Application Fee	Examination Fee	Annual Recertification Fee or Dues
Toxicology	Forensic Toxicology	American Board of Forensic Toxicology	www.abft.org	FSAB	410	$150	N/A	$100
Anthropology	Forensic Anthropology	American Board of Forensic Anthropology	www.theabfa.org	FSAB	79	$250	$300	$100 Annual Dues only
	Forensic Psychology	American Board of Forensic Psychology	www.abfp.com		299; (Outside of FSSP)	$125	$450	N/A
	Forensic Psychiatry	American College of Forensic Psychiatry	www.forensicpsychonline.com		Outside FSSPs	N/A	N/A	N/A
	Forensic Psychiatry	American Board of Psychiatry and Neurology	www.abpn.com	American Board of Medical Specialties	Outside FSSPs	$700	$2,300	$150
	Forensic Nursing	International Association of Forensic Nurses	www.forensicnurses.org		1500+; (Outside of FSSP)	$275/IAFN Member; $400/Non-Member	$400/IAFN Member; $525/Non-Member	$116
Odontology	Odontology—Bite Mark	American Board of Forensic Odontology	www.abfo.org	FSAB	160	$400	$1,000	$230

Figure 2.6 (Continued)

50 Guide to the ABC Biology Exam

the Board of Forensic Document Examiners (BFDE), and the International Association of Computer Investigative Specialists (IACIS).

A primary goal of forensic certification is to bind analysts to a code of ethics. Ethics are of paramount importance in forensic science because the impact of forensic evidence on the criminal justice system is significant. If one piece of evidence is mishandled, manipulated, or misinterpreted, a person's life could be destroyed or justice can go unserved. While the accrediting body defines a list of guiding principles, it's not likely the accrediting body would revoke the accreditation of an entire laboratory based on the actions of one analyst. For example, if an analyst testifies improperly about their training or education, and the laboratory discovers this, they could take swift action in terminating the analyst who is then free to seek employment at some other laboratory where the indiscretions may not be disclosed or discovered. In this instance, certification serves to govern the actions of the individual analysts, and hold them accountable for their conduct in the laboratory and in the courtroom. If the certifying body revokes an analysts certification for ethical violations, moving across the country may not be enough to conceal those indiscretions. Every forensic scientist should be licensed or certified, just as is required of lawyers, doctors and even hairdressers if for no other reason than ethical code enforcement.

2.1.4 Standardization

Accreditation and certification have been major components in the national discussion of forensic science standardization in the United States for many years. In 2009, the National Academy of Sciences published the report, "Strengthening Forensic Science in the United States: A Path Forward." The report recommended mandatory certification of forensic analysts and mandatory accreditation of forensic laboratories and service providers. In 2013, the National Commission on Forensic Science was established to help formulate and recommend standards for the forensic community. The Commission has made both formal and informal recommendations to the attorney general regarding the need for both mandatory certification and mandatory accreditation, cementing the importance of such measures for forensic practitioners and service providers.

Since standardization is the goal of forensic accreditation and certification, organizations that develop standards, such as the American Society for Testing and Materials (ASTM), the American Standards Board (ASB), and the United Nations Office on Drugs and Crime (UNODC) have assumed an important role in setting community standards.

Through standardizing bodies, members of the community are able to meet and reach consensus standards that govern the practice area. In forensic science, the technical working groups (TWGs) began as the first means of setting consensus standards in the forensic disciplines. Later re-named scientific working groups (SWGs), some of these bodies still exist and continue to work toward improvement and standardization of their forensic discipline. In 2014, The National Institute for Standards and Technology (NIST) created the Organization of Scientific Area Committees (OSAC) for forensic science. These committees, intended to replace the SWGs, were established to coordinate development of standards and guidelines and to improve quality and consistency of work in the forensic science community.

Figure 2.7 The National Institute of Standards and Technology is one of the nation's oldest physical science laboratories. NIST provides technology, measurements, and standards that support all scientific projects and creations. (Source: NIST; National Commission on Forensic Science.)

2.2 Quality Assurance/Quality Control Application

As part of any good quality management system, there are checks and balances, and in forensic science the summation of these checks and balances is called quality control. Quality control measures are the means by which forensic scientists monitor the quality of the ongoing work at a lab. These checks are implemented in a variety of ways. Each day, quality control tests designed to produce an expected result are performed. In forensic biology, we have *positive controls*, which should produce an expected positive result when tested with certain reagents or equipment. For example, in forensic DNA, at the amplification step, an analyst will process a sample of DNA that contains a known DNA profile, and this should produce that known profile during the amplification in order to demonstrate that the chemical reagents and equipment are functioning properly. Conversely, *negative controls* should not react with testing chemicals or produce results. An example of a negative control for forensic biology might be testing blood screening chemicals on a swab where no blood is present to see whether a reaction occurs when it should not be occurring. If a control run in the laboratory did not produce its expected result, the lab would consider that result non-compliant data, and there would be a process or procedure for documenting and investigating that anomaly. There are some accreditation standards that direct the investigation of certain types of non-compliant data, but most often a robust quality program put in place by the laboratory will guide lab analysts on how to properly document, investigate, and remediate any non-conformity.

The review and documentation of quality control procedures in the laboratory is very important. It's not enough in the field of forensic science to simply say something was performed in accordance with standards and guidelines; steps must be taken to document and demonstrate that compliance. Quality control tests are repeated and documented on a regular schedule, and records are kept of the person who performed the tests and of the date and time they were performed, so that this information can be demonstrated to an auditor or judge. The Quality Assurance Manager of the lab is responsible for monitoring this documentation and ensuring all these quality control tests are completed and documented properly.

One of the most important functions of a quality assurance program is to oversee the implementation of new processes, procedures, and technology in the lab. These new technologies are introduced after intense study and comparison, called a validation. A "validation involves performing laboratory tests to verify that a particular instrument, software program, or measurement technique is working properly. These validation experiments typically examine precision, accuracy, and sensitivity, which all play a factor on the three Rs of measurements: reliability, reproducibility, and robustness" (https://www.thermofisher.com/us/en/home/industrial/forensics/human-identification/forensic-dna-analysis/forensic-dna-data-interpretation/valid-validation-software/what-is-validation.html).

There are two main types of validations: developmental and internal. A developmental validation is performed when a novel technology is being developed and applied to an area of forensic science for the very first time. This validation determines the reliability, reproducibility, sensitivity, and precision of the results. All aspects of the instrument, process, or technology are studied. This is a detailed process meant to discover all of the strengths and weaknesses. Scientists perform a series of tests meant to demonstrate how the technology works, including tests to show what reagents should be used and what the instrument's settings should be for optimal results. Developmental validation will determine the laboratory protocol for use, and it will be published by the developer for review by other members of the field who may be interested in instituting this technology in their own laboratories. This developmental validation will be viewed as a roadmap for laboratories when they decide to investigate the novel technology for implementation with their own internal validation.

When a laboratory aims to implement a technology not currently in use in their lab (not novel, but new to the lab), the goal is to improve some process or introduce some new scientific advancement. In the evaluation of the new technology, the laboratory needs to determine if it functions as good (or better) than what they currently have in place. This is done by performing a series of tests on each method and comparing the two.

Quality Assurance and Quality Control

How did they perform? Are they equally sensitive? Do they produce comparable results? What is the range of conditions where a reliable result can be obtained? Even though a technology is new to a particular laboratory, a developmental validation has likely already been performed. The purpose of performing the internal validation is to see how the instrument performs in *that* laboratory, with *those* technicians, with *that* instrumentation. It aims to determine how the technology works in their hands. Some necessary parts of an internal validation might involve processing casework-type samples, single-source samples, mixed samples, and samples that test the upper and lower sensitivity range of that technology. SWGDAM defines the following important quality control and quality assurance validation terms:

- Sensitivity: the extent to which the testing method can detect extremely small or dilute amounts of the substance of interest. For example, the range of DNA quantities able to produce reliable typing results must cover the range of DNA concentrations encountered in the samples to be analyzed using the technique.
- Specificity: indicates to what extent the test is likely to give positive results in the case that the substance of interest is present and negative results in case of absence of the substance of interest.
- Repeatability: the variation in measurements obtained when one person measures the same unit with the same measuring equipment.
- Reproducibility: the variation in average (mean) measurements obtained when two or more people measure the same parts or items using the same measuring technique.
- Linearity: the range over which the output signal strength varies in direct proportion to the input signal.
- Limit of detection (LoD): the lowest concentration at which detection is feasible.
- Limit of quantitation (LoQ): the lowest concentration at which not only can a substance be reliably detected, but some predefined goals for bias and imprecision are also met. The LoQ may be equivalent to the LoD or it could have a much higher concentration.
- Limits of reporting: the lowest and highest concentrations that can reliably be reported by the laboratory based on internal and external validations.
- Confidence interval: these are estimates of an interval that may contain the data produced. Confidence intervals are associated with a confidence level, such as 95%, selected by the user. That means that if the same population is sampled on numerous occasions and interval estimates are made on each occasion, the resulting intervals

would bracket the true population parameter in approximately 95% of the cases.

- Confidence limit: the numbers at the upper and lower limit of a confidence interval.

By knowing and understanding these criteria, the laboratory is able to ensure the results from samples being tested are reliable if they fall within the range defined by that laboratory in the validation. A good example for demonstrating why all these criteria are important is the DNA amplification step. Amplification of DNA using Polymerase Chain Reaction (PCR) methods works best within an optimal range. If too little DNA is input, you may obtain partial results or no results at all. If too much DNA is input, amplification and detection will be affected with artifacts that make interpretation of the DNA profile difficult. By knowing the range where reliable results can be obtained, the analyst can make informed decisions about the best use of the sample while guarding against waste.

When instrumentation is validated, the optimal range of operation is established. The instrument is then monitored by the laboratory personnel to confirm that the instrument remains within that range, ensuring reliable results. Calibration procedures are performed routinely in order to maintain the quality of results produced and ensure that the instrument continues to operate in the optimal range. An example of instrumentation that requires constant monitoring and maintenance is an analytical balance. If someone in the laboratory needs to weigh a chemical to create a reagent, it is imperative that the measurement is accurate or else the reagent may not function as intended. With balances and scales, calibration is a regular function of their maintenance to ensure accurate weights and measurements. Standard reference material, such as NIST traceable standard weights, is often used to ensure the readings are in concordance with what is expected. On its website, NIST describes its role in establishing traceability of measurement for forensic quality control and quality assurance:

> NIST is responsible for developing, maintaining and disseminating national standards for the basic measurement quantities, and for many derived measurement quantities. NIST is also responsible for assessing the measurement uncertainties associated with the values assigned to these measurement standards. As such, the concept of measurement traceability is central to NIST's mission. NIST's customers frequently ask questions about traceability and about NIST's role in traceability. It is not always obvious what NIST's role is in helping other organizations establish traceability of their measurement results to standards developed and maintained by NIST. The primary purpose of the NIST Policy on Metrological Traceability is to state the NIST role with respect to traceability. The Policy presents the definition of measurement

Quality Assurance and Quality Control

traceability used by NIST, and clarifies the roles of NIST and others in achieving traceability of measurement results for measurements both internal and external to NIST.

Here NIST describes the responsibility of each laboratory to establish as part of its quality program uncertainties of measurement associated with its methods. Continuing on with the balance example, scales and balances are subject to a certain level of uncertainty, which is usually expressed in the manufacturer's product information. The estimate of uncertainty can be very important when calculating the overall uncertainty of a procedure, and may be necessary when expressing some forensic results, such as drug weights. The laboratory quality program would also have procedures in place for estimating and reporting uncertainty of measurement associated with a method or instrument.

If, during the administration of the quality control procedures, it was determined that an instrument that was out of range was used in the course of testing samples, a series of documentary measures must be taken to address a possible non-conformance. A corrective action may need to be issued, which involves the formal investigation and documentation of the incident. The root cause of the non-conformance must be determined and, if necessary, preventative action must be taken to inform the laboratory of what has occurred and the steps put in place to ensure it doesn't happen again in the future. Corrective and preventative actions formally address any steps that must be taken to rectify the non-conformance and to guard against future repetition of the event.

Another vital role of the quality program is the administration of proficiency tests. Proficiency tests are required under accreditation guidelines to be administered to laboratory personnel periodically. While competency tests are administered at the completion of a training program to demonstrate an analyst's initial capability to perform testing in a certain area, proficiency testing is intended to demonstrate an analyst's ongoing ability to perform testing in that area. Proficiency tests contain mock samples that are processed using the same protocols and controls as casework samples would be processed. The results of the tests are known by the test creators, and are used to evaluate the ability of the analyst to obtain the correct result.

Some proficiency tests are generated by the quality department for the laboratory analysts—these are referred to as internal proficiency tests. Some tests are created by an independent agency, such as Collaborative Testing Services, and sent to the laboratory for processing. Once those tests are processed and results are obtained, the results are then reported back to the independent agency to determine if they are correct. Proficiency tests coming from an outside agency are referred to as external proficiency tests.

Sometimes an analyst knows they are taking a proficiency test and sometimes they do not. If an analyst is unaware they are being tested, the proficiency test is said to be a blind proficiency test. The results of all proficiency tests are recorded and maintained by the quality program for future reference.

2.3 Document/Data Management

Just as all cases must be documented, and that documentation retained, so too must the quality program to create the proper audit trails for review by accrediting agencies. All the quality control and quality assurance procedures and documents are stored for future reference. Much of this information is now stored electronically within laboratory information management systems (LIMS). Just like case documents, quality documents must be preserved to demonstrate the integrity of the quality system.

2.4 Safety

Maintaining a safe environment is imperative in a forensic setting. Forensic labs handle human body fluids, which can contain any number of infectious diseases. They process evidence using harmful chemicals, exposure to which may have significant impacts on the health of the lab personnel. Specific procedures must exist, as required by the accrediting body, to educate and protect analysts from the biological and chemical hazards involved in forensic work.

2.4.1 Chemical Hygiene

One critical requirement included in the quality assurance program is that of maintaining and documenting the chemical hygiene plan for the laboratory. The laboratory will keep an active list of what chemicals are currently being employed and a database of the safety data sheets (SDS) on those chemicals, which delineate the safety hazards of each chemical.

Laboratory personnel would also ensure all chemicals are labeled with the required information, which includes the name of the chemical, the SDS data, the manufacturer's name and contact information, date and initials of individual who received the chemical, when the chemical was opened, and when the chemical expires. The laboratory might also include whether that chemical was tested by the quality control department for suitability, whether it passed quality control, the date when the QC tests were performed, and the initials of the analyst who performed them. To attempt to standardize how laboratories label chemicals, the Globally Harmonized System of Classification and Labeling of Chemicals (GHSCLC) was developed by the

Quality Assurance and Quality Control

United Nations in 2002 (http://www.unece.org/trans/danger/publi/ghs/histback_e.html).

According to the Occupational Safety and Health Administration (OSHA) 29 CFR 1910.1450, the chemical hygiene plan should include:

- Standard operating procedures for laboratory chemicals
  - Procedures for chemical procurement, receipt, and handling
  - Identification of personnel responsible for laboratory chemicals
  - Chemical inventory
  - Chemical storage
  - Chemical handling
  - Definitions of chemical hazards
    - Toxins, corrosives, allergens, asphyxiants, carcinogens, reproductive/embryo toxins
- Compressed gases
- Radiation protection program
  - Monitoring, exposure, training
- Personal protective equipment
- Methods/routes of contamination
  - Inhalation, absorption, ingestion, injection
- General laboratory work practices
  - Grooming, appropriate attire, working alone, handling of chemicals
- Criteria for the implementation of control measures
- Engineering controls employed in the laboratory
- Employee information and training
- Medical information
  - Specific information covering the "Who, what, when, why, and how"
- Chemical hygiene plan responsibilities
  - Safety officer
  - Safety committee
  - Lab director
  - Supervisors
  - Employees
  - Employer
- Record keeping
- Annual audit
- References and recommended reading

In addition to the chemical hygiene plan, the forensic laboratory must also maintain a hazard communication plan. This is mandatory for any lab that uses chemicals. The hazard communication plan entails documenting delivery and receipt of chemicals, monitoring expiration dates, handling disposal, and ordering.

2.4.2 Universal Precautions

Because forensic laboratories also involve manipulation and testing of human body fluids, procedures and protocols for ensuring universal precautions must be observed by all laboratory personnel. For forensic DNA laboratories, the FBI Quality Assurance Standards require the laboratory safety procedures to reflect plans for chemical hygiene and blood-borne pathogens, as well as documentation that laboratory personnel have undergone training with regard to these plans. Since a forensic analyst will likely not know if a crime scene sample contains an infectious disease or blood-borne pathogen, they should treat every sample as potentially infectious. Forensic analysts wear personal protective equipment such as lab coats, latex or nitrile gloves, eye protection, and face masks, to prevent any biological material from the crime scene sample from contacting their body.

2.4.3 Hazardous Waste

One final concern with regard to safety is the handling of hazardous waste and the potential for chemical spills. Since there are many chemicals in use in the forensic lab that are hazardous to human health, it is important to have procedures for storage and disposal of chemical waste as well as procedures for remediation and notification of chemical spills. Chemical spill kits should be available in the lab as well as procedures and protocols for cleaning up spills and disposing of the waste resulting from them.

Figure 2.8 Chemical spill kits are used for cleaning up chemical spills and disposing of waste from spills. (Source: Shutterstock.com.)

Quality Assurance and Quality Control

Bibliography

Armbruster, D. A., & Pry, T. (2008). Limit of blank, limit of detection and limit of quantitation. *The Clinical Biochemist Reviews*, *29*(Suppl 1), S49–S52.

NIST/SEMATECH e-Handbook of Statistical Methods, http://www.itl.nist.gov/div898/handbook/prc/section1/prc14.htm

29 CFR 1910.1450 National Research Council Recommendations Concerning Chemical Hygiene in Laboratories (Non-Mandatory), Appendix A and Appendix B.

29 CFR 1910.1200 Hazard Communication.

Basic Legal and Scientific Concepts

3

3.1 Legal Decisions

When it comes to admissibility of forensic evidence, there are two main cases that define the standards by which the evidence will be measured. The first case is *Frye v. United States*. This case, established in 1923, states that evidence is admissible if it was developed using scientific techniques or methods determined to be "generally accepted" as reliable by the relevant scientific community. This was the first standard by which forensic evidence was measured, and many in the legal and forensic communities feel the standard is a high one because it precludes new and novel (but still reliable) methods from being admitted as they may not yet have reached the level of "general acceptance."

Since many industries, like forensic science, advance quickly, at times even developing new technologies for specific use in a particular case, the Supreme Court of the United States accepted the case of *Daubert v. Merrell Dow Pharmaceuticals* for review in 1993. In *Daubert*, not only did the court hold that scientific evidence could be admissible if it were generally accepted, but the justices also identified several other criteria by which it could be found reliable without yet having reached the level of general acceptance. Under *Daubert*, the judge acts as the "gatekeeper" and decides if the evidence is reliable and admissible using several factors that are suggested but not required, such as:

- Has the method or technique been tested?
- Has it been subject to peer review by means of publication?
- Is there a known potential rate of error?
- Are there standards governing the technique's operations?
- Has the theory or technique been generally accepted by the relevant scientific community?

The *Daubert* decision incorporates some of the language from the Federal Rule of Evidence 702, which governs the admissibility of expert testimony and states that a witness who is qualified as an expert by knowledge, skill, experience, training, or education may testify in the form of an opinion or otherwise if:

(a) The expert's scientific, technical, or other specialized knowledge will help the trier of fact to understand the evidence or to determine a fact in issue.
(b) The testimony is based on sufficient facts or data.
(c) The testimony is the product of reliable principles and methods.
(d) The expert has reliably applied the principles and methods to the facts of the case.

The decision in *Kumho Tire Co. v. Carmichael* (1999) extended the *Daubert* standard for evaluating admissibility of evidence to experts who may not necessarily be scientists. In the case of *Brady v. Maryland* (1963), the court held that the prosecution is required to disclose any exculpatory evidence to the defense. This means that if the prosecution, in the course of its investigation, comes across evidence that tends to show the defendant is not the person who may have committed the crime, the prosecution has a professional responsibility to turn that evidence over to the defense. If a witness or investigator has a history of lying or has committed perjury, that must also be disclosed. Lastly, in the case of *Melendez-Diaz v. Massachusetts* (2009), the 6th Amendment right to confront witnesses was reaffirmed when the court held that in order to submit forensic evidence to the court, a forensic scientist must appear personally to provide testimony on the evidence, so that the defense is given the opportunity to cross-examine the analyst.[1]

3.2 Legal Terms

Some legal definitions that are important to review are:

- **Chain of custody:** Formal record of who has had possession of the evidence and where it was stored at all times. Typically done electronically now with the use of laboratory information management systems.
- **Discovery:** Pre-trial process where parties to a lawsuit or action furnish each other with the evidence they plan to present at trial through several different means (deposition, motions for production of documents, etc.).
- **Voir dire:** A series of questions can be asked to the jury to ascertain whether they harbor some bias with regard to a case issue and can also be asked of an expert to ascertain qualifications and expert knowledge necessary to present testimony on a particular topic.
- **Duces tecum:** Latin for "bring the evidence," usually included in a court order requiring the presence of the forensic examiner to give testimony and further ordering them to produce any relevant evidence they might have in their possession.
- **Subpoena:** A writ requiring one's presence in court to give testimony.

Legal and Scientific Concepts

Molecular biologists are required to have at least one testimony observed and reviewed each year testimony is provided. This is required by the FBI QAS. It states:

> **Standard 3.2** The laboratory shall maintain and follow a procedure regarding document retention that specifically addresses proficiency tests, corrective action, audits, training records, continuing education, case files, and court testimony monitoring.
>
> **Standard 12.7** The laboratory shall have and follow a program that documents the annual monitoring of the testimony of each analyst.

3.3 Court Testimony

When a forensic scientist appears in court, they should appear neat and professional. Proper personal hygiene should be observed, including trimmed and maintained nails with no chipped polish and neatly trimmed hair and facial hair. No tattoos or piercings should be visible. Dress is business professional—dark suits and collared shirts are appropriate. Depending on location, it may be the custom for women to wear skirt suits. No bare arms should be displayed in the courtroom. A forensic analyst should avoid loud colors and flashy jewelry. Conservative clothes are less likely to offend members of the jury than are wild, ostentatious outfits, even when neat. A forensic analyst should always be conscious of his/her demeanor while in the immediate area of the courthouse. The members of the jury begin assessing credibility and forming opinions from

Figure 3.1 Analyst testifying in court. (Source: Shutterstock.com.)

64 Guide to the ABC Biology Exam

the moment they lay eyes on the witness. An attempt should be made to walk naturally when approaching the witness stand and to avoid overly friendly or cold greetings with either defense or prosecution attorneys.

Courtroom demeanor is something that is developed over time. A forensic analyst should make every attempt to respond to questions without emotion or partiality. It is important to speak slowly, loudly, and clearly enough to be heard by everyone in the room. Since the forensic analyst is an impartial witness, there should be no change in demeanor from direct examination to cross examination, and the tone and volume of one's voice should remain the same throughout the testimony.

While it is permissible for a forensic analyst to refer to their case file notes to help refresh their memory while testifying, it is routine for permission to be sought and granted by the judge before referring to them. And while permission is usually granted by the judge, it is always necessary to review the case file notes prior to testimony. One way to make an impression as a competent witness is to have important case information, such as names of suspects or victims involved, committed to memory. On exiting the courtroom, the witness should wait to be excused by the judge and should leave the courtroom without smiling, speaking, or glaring at anyone.

3.4 Procedural Law

With regard to courtroom procedure, many of the arguments and disputes between prosecution and defense take place during pre-trial hearings. For example, if evidence is being challenged as inadmissible under the *Daubert* standard, there would be a *Daubert* hearing before the trial begins to determine whether or not the evidence will be heard at trial. Hearings can be held at any point during the trial proceeding, and they always take place outside the presence of the jury.

A trial is where both prosecution and defense present their cases to the trier of fact, whether that be a judge or a jury. In a standard trial, there would be opening statements by both parties, followed by the presentation of the prosecution case, evidence, and witnesses. In order for forensic evidence to be introduced at trial, a forensic scientist would have to appear to sponsor that evidence. The state is often the first to introduce any forensic evidence. After the presentation of the prosecution case, the defense case, evidence, and witnesses are presented. After each side presents their case, closing arguments are delivered. After each side has rested, the case will go to the judge or jury for decision. Once a decision or verdict has been rendered, sentencing may or may not follow depending on that decision.

During a criminal trial, the burden of proof is on the state. That is why it is said that a person is "innocent until proven guilty": a defendant need not say anything if the state has not met the burden of showing that the defendant

Legal and Scientific Concepts

committed the crime. The burden in a criminal case is guilt beyond a reasonable doubt, which is a high standard. This differs from the burden in civil cases, which is "by a preponderance of the evidence."

Appeals usually take place after the initial trial. An appeal is when the decisions and proceedings of the case are examined to ensure that everything took place in accordance with the law. Contrary to popular belief, this is not a re-trial of the case facts, but rather an examination of the decisions regarding objections, jury instructions, and other legal processes to ensure the ultimate decision of guilt or innocence was reached fairly. If an error of law or procedure is found during appeal, there are a number of outcomes that can result, such as an overturned verdict or the granting of a new trial.

When a forensic scientist is called to court to provide testimony, the court will issue a formal subpoena. Subpoena is Latin for "under penalty." This is most often a court or government order compelling testimony by a witness under some penalty for failure to appear before the court. There are two types of subpoenas: *duces tecum*, meaning "bring the evidence" and *ad testificandum,* ordering testimony. Failure to honor a subpoena can result in a warrant issued for the person's arrest as well as criminal fines and penalties.

A forensic scientist may also have to complete an affidavit, or sworn statement of facts. These are typically submitted before trial by those who are expected to appear and give testimony in court. Affidavits can outline the facts that constitute the probable cause to issue a warrant for search or arrest. In order for these witnesses to appear and testify to the sworn statement in open court, they need to be subpoenaed.

As was mentioned in the discussion of *Daubert,* there are rules that exist to govern the admission of evidence in federal criminal court. Many states have adopted these rules for their state courts as well. For forensic purposes, the most important of these are the rules that deal with expert witness testimony, Rules 701 and 702:

- Rule 701 states that lay witnesses can testify to those things they perceived themselves. This means things they saw, smelled, heard, or observed, but not anything based on technical, scientific, or specialized knowledge.
- Rule 702 states that if scientific or specialized knowledge is necessary, only experts qualified by training and experience in the area can give opinions in court.

3.5 General Science Terms and Principles

The scientific method is a method of inquiry designed to minimize bias and influence from external sources. All scientific analysis is based on this method, including the tests and procedures used in forensic science. There are five basic steps to the scientific method of inquiry:

1. Observation.
2. Formulation of a hypothesis, usually an "if, then" statement that involves an educated guess about what the results of the inquiry will be.
3. Making additional observations and testing the hypothesis.
4. Analyzing the data generated during tests.
5. State conclusions based on analysis of the data—is the stated hypothesis supported or rejected?

3.5.1 General Chemistry Concepts

For the exam, there are some basic concepts from each scientific field which are important to be familiar with. For general chemistry, just as it was noted that there are some bodies charged with standardization of terms and definitions for certain forensic disciplines, so too do they exist for chemistry. One of those bodies is the International Union of Pure and Applied Chemistry, or IUPAC. According to their website, "The International Union of Pure and Applied Chemistry (IUPAC) is the world authority on chemical nomenclature and terminology, including the naming of new elements in the periodic table; on standardized methods for measurement; and on atomic weights, and many other critically-evaluated data."

Stemming from organic chemistry, there may be questions on the exam dealing with the types of hydrocarbons and their derivatives. These compounds make up the bulk of living matter and play an important role in forensic science. Hydrocarbons are carbon- and hydrogen-containing compounds that are classified as either saturated or unsaturated.

Table 3.1 Table Showing Alkane Names for Chain Lengths up to Ten Carbons

Name	Number of C Atoms	Molecular Formula C_nH_{2n+2}	Condensed Structural Formula
Methane	1	CH_4	CH_4
Ethane	2	C_2H_6	CH_3CH_3
Propane	3	C_3H_8	$CH_3CH_2CH_3$
Butane	4	C_4H_{10}	$CH_3CH_2 CH_2CH_3$
Pentane	5	C_5H_{12}	$CH_3CH_2 CH_2 CH_2CH_3$
Hexane	6	C_6H_{14}	$CH_3CH_2 CH_2 CH_2 CH_2CH_3$
Heptane	7	C_7H_{16}	$CH_3CH_2 CH_2 CH_2 CH_2 CH_2CH_3$
Octane	8	C_8H_{18}	$CH_3CH_2 CH_2 CH_2 CH_2 CH_2 CH_2CH_3$
Nonane	9	C_9H_{20}	$CH_3CH_2 CH_2 CH_2 CH_2 CH_2 CH_2 CH_2CH_3$
Decane	10	$C_{10}H_{22}$	$CH_3CH_2 CH_2 CH_2 CH_2 CH_2 CH_2 CH_2 CH_2CH_3$

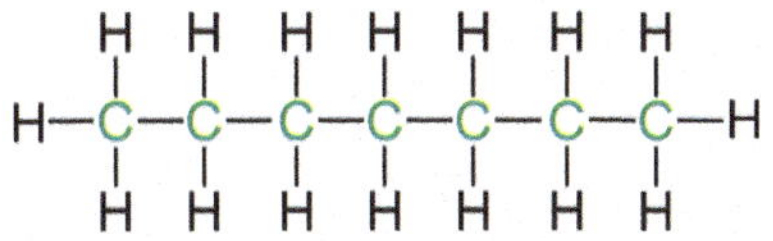

Figure 3.2 Saturated hydrocarbon—no double bonds exist in the molecule. (Image retrieved from Wikimedia, used per Creative Commons 3.0, user Dan1gia2.)

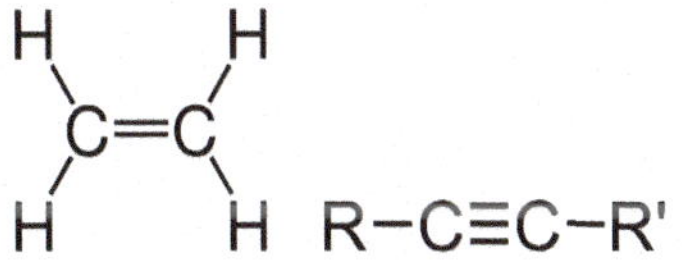

Figure 3.3 Unsaturated hydrocarbons. These molecules are composed with some carbon-to-carbon (a) double or (b) triple bonds. (Image (a) retrieved from Wikimedia, used per Creative Commons 3.0, user Poyrax 72.)

Alkanes and cycloalkanes are saturated hydrocarbons. These are compounds containing only carbon–carbon single bonds. If a compound is said to be saturated, it is because the carbon has the maximum number of hydrogen atoms bonded to it. All alkanes have the same general formula, C_nH_{2n+2}: for example, propane is a three-carbon hydrocarbon and its formula is C_3H_8. Cycloalkanes are hydrocarbons where the carbon atoms are joined to form a ring. Cycloalkanes have the general formula C_nH_{2n}. Isoalkanes contain a branched-chain within the hydrocarbon molecule. There are specific IUPAC rules that describe the naming of each of these compounds.

The unsaturated hydrocarbons are the alkenes, alkynes and arenes (aromatic compounds). These are compounds that contain at least one carbon–carbon double or triple bond. Unsaturated hydrocarbons are generally more reactive given the fact that they are not fully saturated, with the exception of benzene, which is relatively unreactive. Alkenes are hydrocarbons that contain at least one carbon–carbon double bond. Alkynes are hydrocarbons that contain at least one carbon–carbon triple bond. Arenes, also called aromatic hydrocarbons, are compounds that contain a benzene ring. A benzene ring is a six-member carbon ring with alternating single and double carbon–carbon bonds, and is the simplest aromatic compound with the formula C_6H_6.

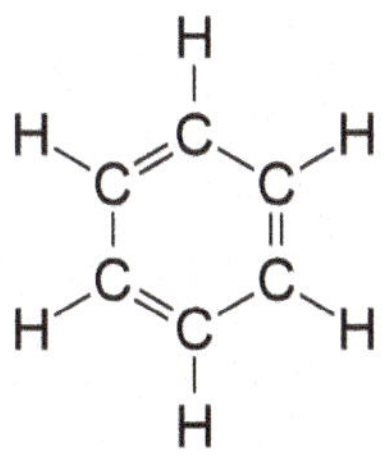

Figure 3.4 A benzene ring is a six-carbon ring with alternating single and double carbon-to-carbon bonds. (Image retrieved from Wikimedia, used per Creative Commons 3.0, user cacycle.)

Figure 3.5 Periodic table of the elements. (Image retrieved from Wikimedia, used per Creative Commons 4.0 International, user OpenStax Anatomy and Physiology.)

Legal and Scientific Concepts

The atomic number is the number of protons in the nucleus of an atom. An element is a substance whose atoms all have the same atomic number. The mass number is the total number of protons and neutrons in a nucleus. Molecular weight is the sum of the atomic weights of the atoms in a molecule. Isotopes are atoms whose nuclei have the same atomic number but different mass numbers, meaning that different isotopes of an atom have the same number of protons but a different number of neutrons.

The periodic table is a list of all the elements according to their atomic size, orbital energy, and effective nuclear charge. It is the organizational framework of chemistry.

Elemental composition is the chemical makeup of a particular given compound. Elements are made up of only one type of atom. Compounds are substances formed when two or more elements are combined in a defined ratio.

Figure 3.6 Ionic bond between sodium and chlorine. In this type of bond, the electron is donated from the sodium atom to the chlorine atom to form the bond.

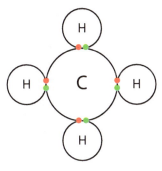

Figure 3.7 Covalent bonds between carbon and hydrogen.

An Arrhenius acid is a substance that dissociates in water to form H^+ ions. An Arrhenius base is a substance that dissociates in water to form OH^- ions. A Brønsted-Lowry acid is a substance that donates a proton (H^+). A Brønsted-Lowry base is a substance that accepts a proton (H^+).

There are several types of bonds that occur in chemistry. One type is the ionic bond, where electrons are transferred from one atom to another. Another is the covalent bond, where electrons are shared between atoms.

Ionic bonds are stronger in nature than covalent bonds. Hydrogen bonds are a weak form of covalent bonds, in which hydrogen shares its electron with nitrogen, oxygen, or fluorine. The weakest bond is made by Van der Waals forces. These forces are created by the natural movement of electrons to one end of the molecule, causing it to be temporarily attracted to another molecule that has experienced a shift of electrons. This is also known as dipole-dipole interaction or London dispersion forces.

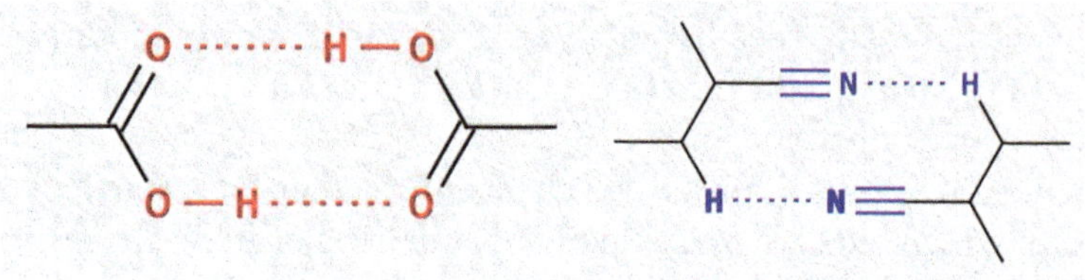

Figure 3.8 Hydrogen bonds, depicted as dashed lines, act as an intermolecular force to hold molecules together.

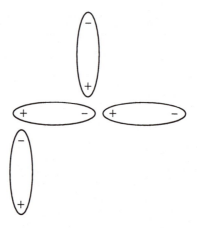

Figure 3.9 General idea behind Van der Waals forces. The more positive end of a molecule will draw toward the more negative end of another molecule.

Stereoisomers are compounds with the same molecular formula that differ only in the spatial arrangement of their atoms. Optical isomers are a type of stereoisomers that are non-superimposable mirror images of one another. Enantiomers are two molecules that are mirror images of the other. Thus, each optical isomer in a pair is an enantiomer.

Forensic science requires a basic understanding of stoichiometry, or the relationship between quantities of materials in chemical reactions. Whether it is for preparation of reagents, or an equation describing a color reaction between screening chemicals and body fluid, it's important to know how to understand and balance chemical equations. Without a basic

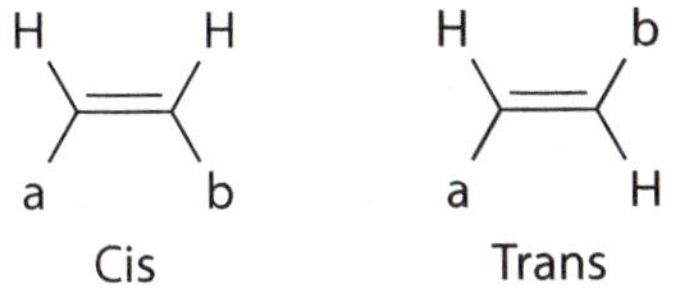

Legal and Scientific Concepts

Figure 3.10 Geometric isomers are two molecules with the same empirical formula, but different arrangement of atoms.

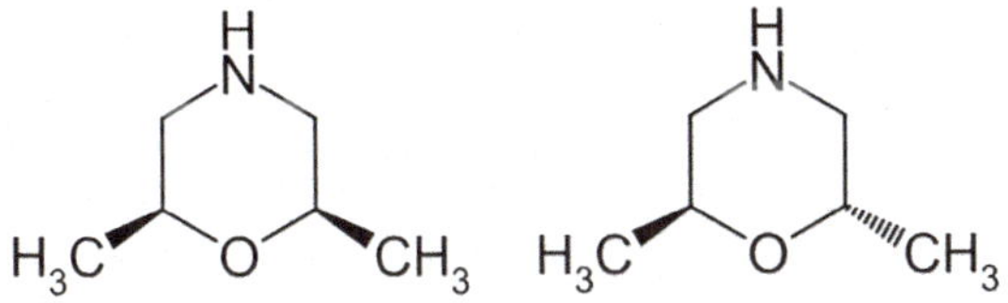

Figure 3.11 Stereoisomers are two molecules that differ only in the spatial arrangement of their atoms.

understanding of this, it makes it impossible to accurately describe the kinds of chemical reactions taking place as a part of forensic testing or identification.

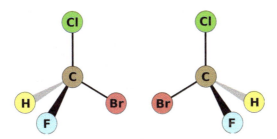

Figure 3.12 Enantiomers are molecules that are mirror images of one other, but are not superimposable. The carbons on these molecules are chiral, since they are attached to four different groups.

3.5.2 General Biology Concepts

From cell biology, there are two types of cells of highest concern in forensic science: animal cells (with which we deal most often) and plant cells. The structure of these cells varies slightly, but they have many of the same components. Each of these cells has a nucleus, where all the genetic material is contained. They both contain cytoplasm, which contains common organelles such as the Golgi apparatus, endoplasmic reticulum, ribosomes, and mitochondria. The cell wall structure of an animal cell is less rigid than that of a plant cell, and the plant cell contains chloroplasts.

A third type of cell, the bacterium, is becoming more and more important in forensic molecular biology with the advent of bacterial profiles and the like. These cells can have different shapes (rod, sphere, twist) and lack a nucleus as well as many of the other commonly encountered organelles. Bacteria contain DNA and ribosomes in the cytoplasm, and also exhibit a cell wall.

From genetics, principles of inheritance were first characterized by Gregor Mendel in the nineteenth century when he described his experiences growing and fertilizing pea plants (discussed in depth in Chapter 6). Since then, human inheritance has been studied to describe the principles of genetic reproduction. Our basic understanding of how genetic material comes together to form the blueprint of a human being advances and grows every day. This study informs forensic molecular biology and the use of biological material in human identification on the most basic level. For example, a person's forensic DNA profile is inherited from their mother and their father in equal parts. For the purposes of forensic DNA typing currently available, identical twins will have the same profile because they share the same genetic makeup, having been born from the division of one cell. While it is expected that the DNA from related individuals will have more genetic similarities than that of unrelated individuals, it is not expected that any two people who are not identical twins will have the same forensic DNA profile. The most important thing to keep in mind about genetics and the way cells reproduce is that it all happens independently, which is a topic we will discuss as we delve into the forensic biology concepts in more detail.

The unique characteristics of body fluids, including the proteins and enzymes contained in each fluid, play an important role in forensic biology. The chemical composition of each fluid, the types of cells that can be identified as components of each fluid, and the types of biochemical reactions that can be used to identify them all play a fundamental role in forensic biology and even formed the basis for the technologies, such as forensic DNA analysis, that exist today.

3.5.3 General Physics Concepts

From physics, the concept of energy has been applied to forensic science. Energy is the ability of a body to do work. Many of the reactions that take place as part of forensic testing require energy to begin and complete the test. Different types of energy used in forensic science include chemical energy, which drives the oxidation reaction in many color tests, such as those that indicate the presence of blood, electrical energy, such as the electrokinetic injection that forces the injection of DNA molecules into the capillary of a genetic analyzer using capillary electrophoresis, or thermal energy, which is applied to cells in a biological sample to lyse them and release any DNA

Legal and Scientific Concepts 73

present. Force, a concept related to energy, is a push or pull on an object. Force is often used to describe energy as an attribute of physical action.

The electromagnetic spectrum is utilized in forensic science because there are many types of tests and instruments that rely on different types of visible and invisible light. The electromagnetic spectrum is the range of wavelengths or frequencies over which electromagnetic radiation extends. There are illegal drug tests that use infrared light, and there are chemical and biological screening tests that rely on color-change reactions of visible light. There are also processes and reagents that rely on differences in wavelength emission and fluorescence. Many forensic tests are deeply rooted in these physics concepts. Likewise, an Alternative Light Source (ALS) is often used when screening evidence to assist in visualizing body fluids that may not be visible with the naked eye. The ALS operates by emitting different wavelengths of light, some visible and others invisible, which certain fluids and materials will refract to a different wavelength that allows it to stand out on the item.

3.5.4 General Physiology Concepts

Forensics borrows from anatomy and physiology in that an analyst must have some basic knowledge of the human body for information about biological processes. Without foundational knowledge of the human reproductive system, where body fluids can be found in the body, what the roles of certain body fluids are in the human body and the types of reactions they're involved in, a forensic scientist can never fully understand how a body fluid can be used as a piece of a forensic investigation. Consider the toxicology involved in blood alcohol testing. It requires knowledge of the human respiratory system for breath testing, and knowledge of metabolic systems for digestion, oxidation, and excretion of alcohol in the human body to know where to look for evidence of intoxication. This knowledge is vital to the discovery and establishment of probative value.

3.5.5 General Statistics Concepts

General statistics have played a role in forensic molecular biology for some time and have been used to give weight to blood protein profiles and now DNA profiles. But never has the field of general statistics been called to play a more prominent role in the overall field of forensic science as it has in the last five or ten years. After the criticisms faced by many disciplines in the NAS Report and other government-funded evaluations of the field, there have been strong efforts to use statistics in order to give weight to "match" statements in disciplines such as latent print comparison and firearms analysis, which have been difficult to quantify. Forensic scientists are being asked to revisit whatever training and education they have in statistics, and to undertake additional training to understand new concepts and calculations that will soon be a part of routine casework for the forensic analyst.

Here, some basic definitions have been included as a refresher in statistics. The mean is simply the average of a dataset. In a normal Gaussian distribution, or bell curve, the mean is the peak of the bell. The median is the middle number in any given list of obtained values or results. When results are obtained and sorted from highest to lowest, the value that lies in the middle is known as the median. When the obtained values or results are in list form, the number that appears most frequently is the mode. An easy way to remember this is: MOst frequent = MOde.

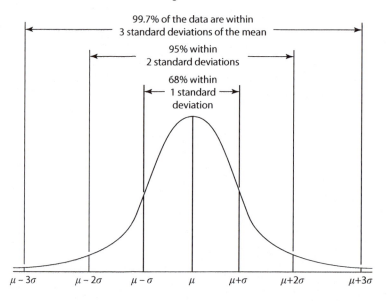

Figure 3.13 A depiction of a normal Gaussian distribution curve showing the percentage of the results that fall within each standard deviation of the mean. (Image retrieved from Wikimedia, used per Creative Commons 4.0 International, user Dan Kernler.)

Standard deviation is the average deviation from the mean. This measures the spread of the values or results. Standard deviation is usually rounded to two significant figures and is an indicator of how tightly grouped the data points are. A small standard deviation means the data points are very close together, thus there is little variability in the data. A large standard deviation means the data points are spread over a larger range and there is greater variability. This is important in forensic science in areas such as metrology (measurement) and validation. For example, when a DNA laboratory endeavors to set thresholds for reporting data on a new instrument, they run a series of tests to obtain data points. Based on a review of the data generated, the laboratory will define a range where the data shows

Legal and Scientific Concepts

reliable results. They may also decide to report results within a certain number of standard deviations to ensure that the results they are reporting fall within the reliable range. In a normal distribution, expanding the range plus or minus one standard deviation from the mean will include approximately 68% of the observed data. Extending out two standard deviations (in either direction) will encompass approximately 95% of the observed data. Extending out three standard deviations (in either direction) will encompass approximately 99% of the observed data, and so the laboratory can feel confident that 99% of their results will be called using that threshold.

Bayes' theorem, named after Reverend Thomas Bayes, describes the probability of an event based on prior knowledge of conditions that might be related to the event. One of the many applications of Bayes' theorem is Bayesian inference, a particular approach to statistical inference. When applied, the probabilities involved in Bayes' theorem may have different probability interpretations. With the Bayesian probability interpretation the theorem expresses how a degree of belief, expressed as a probability, should rationally change to account for availability of related evidence. Bayesian inference is fundamental to Bayesian statistics. The likelihood ratio as applied to forensic DNA profiles relies on Bayesian theory.

One final point to consider about statistics is the importance of population or sample characteristics and how they can impact results. Suzanne Bell writes in her textbook *Forensic Chemistry*, "As the number of measurements of the population increases, the average value approaches the true value. The goal of any sampling plan is two-fold: first, to insure that 'n' is sufficiently large to appropriately represent the characteristics of the parent population; and second, to assign quantitative, realistic and reliable estimates of the uncertainty that is inevitable when only a portion of the parent population is studied."[2] An example of this from forensic biology is the allele frequencies assigned to the different alleles in a DNA profile. In the 2nd revision of the NRC report, NRC II, the authors address the importance of database characteristics as they apply to the frequencies applied to genetic traits in DNA profiles: "Match probabilities are estimated from a database, and such calculations are subject to uncertainties. The accuracy of the estimate will depend on the genetic model, the actual allele frequencies, and the size of the database." Additionally, as many laboratory units are reported using the metric system, the forensic analyst should be familiar with the units of measurements that apply to the processes they conduct.

3.5.6 Logic

Finally, for any forensic scientist, critical thinking is an important skill. Taking the facts of a case and evidence recovered, analyzing the information and developing a testing plan, re-evaluating the plan based on results

METRIC CONVERSION CHART

when you know	multiply by	to find	when you know	multiply by	to find
length			**mass and weight**		
millimeters	0.04	inches	grams	0.035	ounce
centimeters	0.39	inches	grams	0.032	ounce (apoth.)
meters	3.28	feet	kilograms	2.20	pounds
meters	1.09	yards	kilograms	2.68	pounds (apoth.)
kilometers	0.62	miles	tons (1,000 kg)	1.10	short tons
inches	25.40	millimeters	ounces	28.35	grams
inches	2.54	centimeters	ounces (apoth.)	31.10	grams
feet	30.48	centimeters	pounds	0.45	kilograms
yards	0.91	meters	pounds (apoth.)	0.37	kilograms
miles	1.61	kilometers	short tons (2,000 lb)	0.91	metric tons
speed			**temperature**		
miles per hour	1.61	kilometers per hour	degrees Fahrenheit	$(°F - 32) \div 1.8$	degrees Celsius
kilometers per hour	0.62	miles per hour	degrees Celsius	$(°C \quad 1.8) + 32$	degrees Fahrenheit
volume			**metric prefixes**		
milliliters	0.20	teaspoons	prefix	symbol	factor
milliliters	0.07	tablespoons			
milliliters	0.03	fluid ounces	exa-	E	10^{18} = 1,000,000,000,000,000,000
liters	4.23	cups	peta-	P	10^{15} = 1,000,000,000,000,000
liters	2.11	pints	tera-	T	10^{12} = 1,000,000,000,000
liters	1.06	quarts	giga-	G	10^{9} = 1,000,000,000
liters	0.26	gallons	mega-	M	10^{6} = 1,000,000
cubic meters	35.31	cubic feet	kilo-	k	10^{3} = 1,000
cubic meters	1.31	cubic yards	hecto-	h	10^{2} = 100
teaspoons	4.93	milliliters	deca-	da	10 = 10
tablespoons	14.79	milliliters	deci-	d	10^{1} = 0.1
fluid ounces	29.57	milliliters	centi-	c	10^{2} = 0.01
cups	0.24	liters	milli-	m	10^{3} = 0.001
pints	0.47	liters	micro-	μ	10^{6} = 0.000,001
quarts	0.95	liters	nano-	n	10^{9} = 0.000,000,001
gallons	3.79	liters	pico-	p	10^{12} = 0.000,000,000,001
cubic feet	0.03	cubic meters	femto-	f	10^{15} = 0.000,000,000,000,001
cubic yards	0.76	cubic meters	atto-	a	10^{18} = 0.000,000,000,000,000,001

Publisher's Design & Production Services

Figure 3.14 Metric conversion chart showing metric to metric conversions as well as English to metric conversions.

obtained, and troubleshooting are all tasks that require a scientist to think critically. While there will likely be a defined protocol or procedure for each test, the protocols can never be written to address every single scenario. It's vital for a forensic scientist to identify important facts and information and use them as a guide as they examine evidence.

Also related to the logical aspect of forensic science are the two types of reasoning: inductive and deductive. Inductive reasoning makes broad generalizations from specific examples. Conversely, deductive reasoning uses the scientific method to draw specific conclusions from general principles. Unlike in inductive reasoning, which always involves uncertainty, the conclusions from deductive inference are certain provided the premises are true. Scientists use inductive reasoning to formulate hypotheses and theories, and deductive reasoning when applying them to specific situations.

Legal and Scientific Concepts

Notes

1. Saferstein, R. (2018) *Criminalistics: An Introduction to Forensic Science* (12th ed.). Upper Saddle River, N.J.: Pearson.
2. Bell, S. (2014). *Forensic chemistry.* Harlow: Pearson Education International.

Bibliography

http://web.eng.ucsd.edu/~jschulze/misc/spinoff/metric-conversion.htmhttp://math.about.com/od/statistics/a/MeanMedian.htmBell, S. (2013). *Forensic chemistry* (2nd ed.). [Bookshelf Online]. Retrieved from https://bookshelf.vitalsource.com/#/books/9780321849960/

NRC IIVining, W. (2015). *General chemistry* (1st ed.). [Bookshelf Online]. Retrieved from https://bookshelf.vitalsource.com/#/books/9781305459809/

http://www.csun.edu/science/ref/reasoning/deductive_reasoning/index.htmlhttps://www.law.ufl.edu/_pdf/faculty/little/topic8.pdfhttps://supreme.justia.com/cases/federal/us/509/579/case.htmlhttps://www.law.cornell.edu/rules/fre/rule_702https://supreme.justia.com/cases/federal/us/526/137/case.htmlhttps://supreme.justia.com/cases/federal/us/373/83/case.htmlhttps://www.rulesofevidence.org/

Principles and Concepts of Biological Screening

4

4.1 Biological Screening Tests

Forensic serology is the identification of biological material related to criminal cases. This includes testing for body fluids, or swabbing items to collect touch or wearer DNA. Body fluids are identified through a series of biological screening tests. The hallmark of any good screening test is sensitivity. If a fluid is or was present recently, an effective screening test should be able to detect it. Screening tests are generally less *specific* for an individual fluid, but more *sensitive*. They're the first line of detection, often being employed in the field to indicate to investigators where they should sample for biological evidence. The absence or presence of biological fluids at a crime scene can confirm witness accounts and help reconstruct crime events. Biological fluids also serve as important sources of evidence that can link perpetrators to their crimes. Commonly encountered fluids in forensics include blood, semen, saliva, urine, and feces.

Tests to identify body fluids can be simply visual—performed using just the naked eye or an alternate light source (ALS)—or they can involve microscopic or chemical examination. Visual techniques include observation of the fluid in natural, bright, or oblique lighting. An example of this would be visualizing a red-brown stain that is suspected to be blood, or a whitish-yellow crusted stain that is suspected to be semen. For fluids that are less easily observed, enhancement methods must be employed. An alternate light source can be used to visualize fluids that fluoresce under blue light (420–470nm) such as semen, urine, sweat, or saliva. The ALS can also be used to visualize bloodstains on dark items using an infrared light source (800–900nm). Fluorescent and chemical mapping techniques can also be employed, and may include the use of a chemical spray to reveal the location of the body fluid on an item. Lastly, microscopy is an age-old but time-tested technique that can be used to identify the presence of a biological fluid.

4.1.1 Blood

Blood is a liquid connective tissue composed of plasma, erythrocytes (red blood cells), leukocytes (white blood cells), and platelets. Blood is used to transport gases, nutrients, and wastes throughout the body, as well as to regulate body temperature and fluid levels. Plasma is the liquid portion of blood that is primarily water, with a pH close to that of water at 7.35–7.45. It accounts for approximately 55% of blood content. Red blood cells transport gases through the body with the use of hemoglobin. Hemoglobin is a protein that contains four heme, or iron-containing, groups. These heme groups bind to oxygen and carbon dioxide, transporting them to and from the lungs and throughout the body.

Red blood cells contain hundreds of antigens on their surface, some of which determine the individual's blood type (A, B, AB, or O). White blood cells produce immune responses throughout the body. Platelets are tiny cells that form blood clots. When blood begins to clot, the cells separate from a liquid called serum. Serum is the clear yellowish fluid that remains after clotting factors (such as fibrinogen and prothrombin) have been removed by clot formation. White blood cells contain nuclei while red blood cells and platelets do not, therefore the DNA provided from blood samples comes from white blood cells.

4.1.1.1 Blood Typing

The evolution of the forensic biology discipline toward the modern-day DNA typing we have today started with blood typing. Blood typing and protein profiling laid the foundation for modern DNA typing. More than 15 antigen systems have been discovered on red blood cells, but the most important are the A-B-O and Rh systems.

People with type A antigens produce type A blood, and those with type B antigens produce type B blood. Cells with both A and B antigens produce

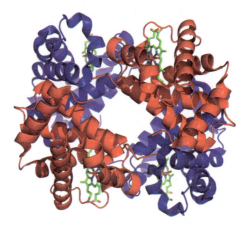

Figure 4.1 Hemoglobin is a protein in red blood cells that binds to oxygen in the lungs and transports it throughout the body. (Image retrieved from Wikimedia, used per Creative Commons 3.0, user Richard Wheeler (Zephyris).)

Biological Screening

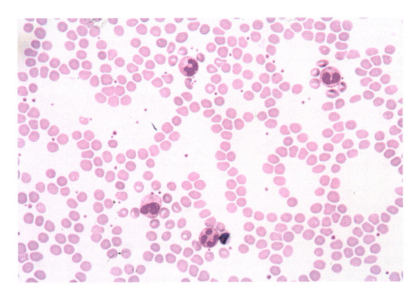

Figure 4.2 Numerous red blood cells and four white blood cells under the microscope. (Source: Shutterstock.com.)

type AB blood, and cells with neither A nor B antigens produce type O blood. The Rh factor, or D antigen, is either present or absent on red blood cells, which is indicated by a positive (+) or negative (−) sign after the blood type. The blood serum contains different antibodies which will each coexist with specific antigens and react with others. The reaction of antigen and antibody creates a phenomenon known as agglutination, which is a clumping together of red blood cells. Blood type A has anti-B antibodies in the serum, which will cause an agglutination reaction with type B blood. Type B blood has anti-A antibodies, which will cause an agglutination reaction with type A blood. Type AB has neither anti-A nor anti-B antibodies, and type O has both anti-A and anti-B antibodies. Thus the identification of the blood type of an unknown blood sample can be made by the combination of blood with known A and B blood and observing the manner of agglutination.

The study of the reactions between antigens and antibodies is known as serology. The most commonly used agglutination test for blood typing in forensic science is the Lattes crust assay. This test was groundbreaking at the time it was discovered because many crime scenes are discovered after the fact, and the blood present has already dried. Previous A-B-O blood typing tests described by Karl Landsteiner worked only with fresh liquid blood. The Lattes crust test could be performed on blood that had already dried, extending the utility of A-B-O blood typing to crime scenes that were not discovered right away.

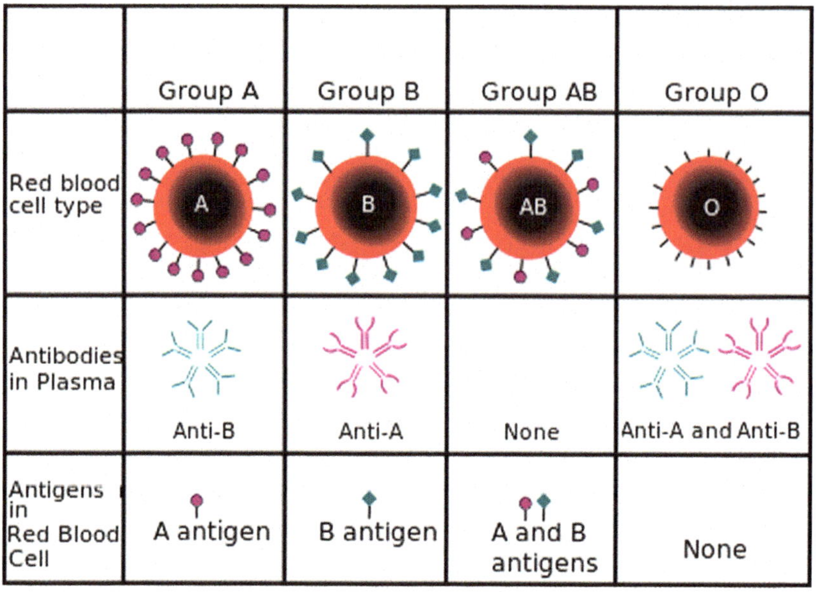

	Group A	Group B	Group AB	Group O
Red blood cell type	A	B	AB	O
Antibodies in Plasma	Anti-B	Anti-A	None	Anti-A and Anti-B
Antigens in Red Blood Cell	A antigen	B antigen	A and B antigens	None

Figure 4.3 Image showing the blood types and their corresponding antibodies and antigens. (Image retrieved from Wikimedia, used per Creative Commons 3.0, user Shahinsahar.)

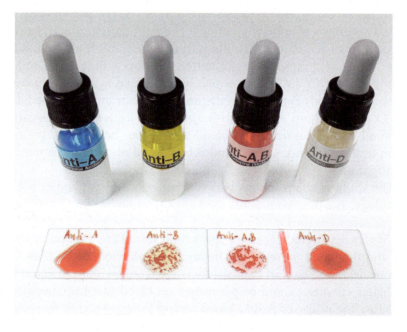

Figure 4.4 Blood group testing by agglutination. (Source: Shutterstock.com.)

Biological Screening

4.1.1.2 Presumptive Tests for Blood

Blood can be detected through visible observation, infrared illumination, or colorimetric, chemiluminescent, and fluorescent assays. Colorimetric assays rely on the peroxidase activity of heme to catalyze an oxidation-reduction reaction with the substrate. These assays include phenolphthalein (Kastle–Meyer), leucomalachite green, benzidine blue, tetramethylbenzidine (TMB), and ortho-tolidine (OT).

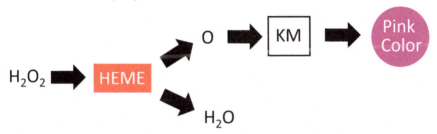

Figure 4.5 A graphic of the reaction occurring during the Kastle–Meyer test for blood. Hydrogen peroxide, when reacted with hemoglobin, breaks down into water and oxygen. The oxygen reacts with the Kastle–Meyer reagent, producing the positive pink color.

Figure 4.6 Leucomalachite green, a presumptive test for blood.

The leading chemiluminescent tests for blood detection are luminol and Bluestar, both of which react with blood to produce a blue-colored light and must be visualized in darkened conditions. The manufacturer reports that Bluestar has a brighter reaction that stays visible longer than luminol, which makes it easier to photograph. The predominant fluorescent test for blood is fluorescein, which makes blood fluoresce a yellow-green color under a wavelength of light in the 425–485nm range. These tests are applied by spraying the reagent directly onto the area where the suspected blood might be present. These reagents do not interfere with subsequent DNA testing, but if only a trace amount of blood was present initially, spraying too much reagent may dilute the sample to a level where the DNA is undetectable. Care must be taken to ensure this does not take place.

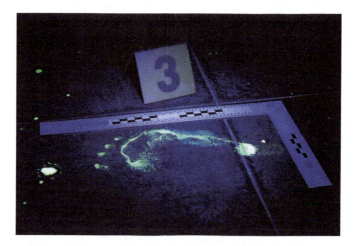

Figure 4.7 A footprint left in blood fluoresces after being sprayed with luminol and visualized under UV light. (Source: Shutterstock.com.)

Because biological screening tests were designed to be sensitive rather than specific, it must be noted that some substances can interfere with presumptive tests for blood, producing false positive results. Some substances that can cause false positives include oxidants such as some metals and cleaning products, plant peroxidases such as horseradish, and strong reductants such as zinc and lithium. Additionally, animal blood will also produce false positive reactions with the presumptive tests for blood.

4.1.1.3 Confirmatory Tests for Blood

In the modern crime lab, the most common confirmatory tests for blood are immunochromatographic cards such as ABAcard HemaTrace® or SERATEC® HemDirect, which can confirm the presence of human blood by using antihuman hemoglobin antibodies. These tests have gained popularity in the forensic community for being inexpensive and easy to use.

Biological Screening

The test itself is a small plastic cartridge that is similar in chemistry and appearance to a pregnancy test. There is a small circular well for addition of the sample where dye-labeled monoclonal antihuman hemoglobin (Hb) antibody is present. An immobilized antihuman-Hb antibody is present in the test zone and an immobilized antiglobulin is present at the control zone. The sample is placed into the well, and if it contains hemoglobin, this will adhere to the dye-labeled monoclonal antihuman-Hb antibody. The dye-labeled antihuman-Hb antibodies travel via diffusion across the nitrocellulose membrane to the test zone. If the hemoglobin has adhered to the dye-labeled antibodies, it will also attach to the immobilized antihuman antibody in the test zone, producing a pink/red line. The more hemoglobin that binds to the immobilized polyclonal antihuman antibody in the test zone, the more dye is concentrated in that area, which produces the pink/red line indicating a positive result. Excess labeled antibodies also migrate to the control zone, where they will adhere to immobilized antiglobulins, producing a pink/red line which indicates the test is working properly.

A negative immunochromatographic assay should display one pink/red line in the control zone to indicate that the test reagents are working properly, but no hemoglobin was present. A positive immunochromatographic assay will display a pink/red line in both the control and test zones. A test should be considered inconclusive if there is no pink/red line in the control area, even if a line is detected in the test zone.

One phenomenon to be aware of when using immunochromatographic tests is the high-dose hook effect. If too much hemoglobin is present in the sample, it can overwhelm the card. Too much free hemoglobin will quickly bind up the dye-labeled monoclonal antihuman antibody present in the sample well. However, excess unbound hemoglobin will also be present. This unbound hemoglobin travels down the membrane faster than bound hemoglobin and will rush toward the immobilized antihuman antibody in the test zone. The unbound hemoglobin can bind to all of the available sites, leaving no available antihuman antibody for dye-labeled conjugates to bind. In this instance, even though there is hemoglobin in the sample, no line will appear in the test zone. The antibody may still bind to the immobilized antiglobulin in the control zone, giving the appearance of a negative result with a single pink/red line in the control zone. Samples that show visible red-brown or straw color should be diluted, and run again to confirm whether high-dose hook may be affecting the result.

RSID, or Rapid Stain Identification, is another common blood confirmatory test that is similar in its chemistries to the immunochromatographic cards described above. It is human specific and does not exhibit a high dose hook effect. RSID-blood tests identify human glycophorin A rather than hemoglobin. Other less commonly utilized confirmatory tests include microcrystal assays including hematin and hemochromogen assays. The hematin

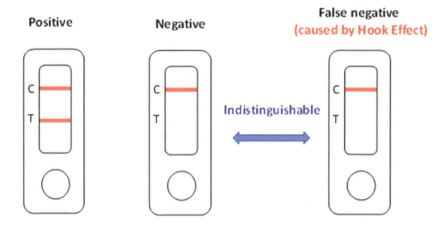

Figure 4.8 High-dose hook effect occurs when too much hemoglobin is introduced into the reaction. The immobilized antibody at the test region becomes saturated with free hemoglobin, preventing the binding of the labeled hemoglobin. When this occurs, no line forms in the test region, producing a false negative result. (Image is reproduced with permission from ANP Technologies, Inc (www.anptinc.com).)

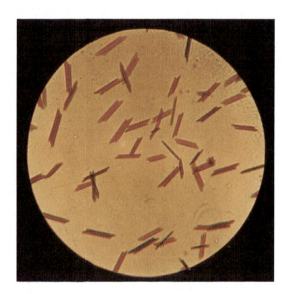

Figure 4.9 Microcrystal assay using the Teichmann method. (From James, S.H., Nordby, J.J., and Bell, S., *Forensic Science: An Introduction to Scientific and Investigative Techniques*, 4th edn., CRC Press, Boca Raton, 2014. With permission.)

Biological Screening

assay, or Teichmann crystal assay, is a process by which the blood is treated with glacial acetic acids and salts, producing a brown-colored ferriprotoporphyrin chloride crystal. The hemochromogen assay, or the Takayama crystal assay, treats the blood with glucose and pyridine, forming the crystal pyridine ferriprotoporphyrin. The hematin assay tends to be the more reliable of the two tests. Blood can also be analyzed using spectrophotometry, demonstrated by a peak at 400–425nm, as well as through chromatographic and electrophoretic methods to identify hemoglobin. Real time PCR (RT-PCR) can also be used in blood identification by detecting mRNA specific to erythrocytes.

4.1.1.4 Blood Species Identification

Sometimes it is relevant to determine the species from which blood at a crime scene originated. If there is a question as to which species a blood sample originated from, the precipitin test can be used. By reacting the sample with human antiserum, human blood will react with the antibodies specific to humans to produce a cloudy ring around the interface of the sample. Animal blood will produce no such ring when interacting with human antiserum.

The Ouchterlony test is a species identification test that relies on using antibodies. In Ouchterlony double diffusion, both antigen and antibody are allowed to diffuse into the gel. Antigens from different species are loaded into wells and the unknown sample is loaded in into a hole punched into the center of the plate. In a positive reaction, a line will develop between the sample and antiserum well.

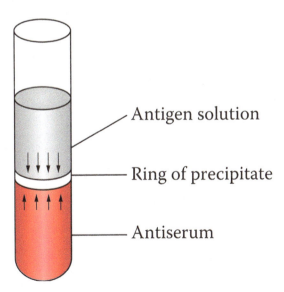

Figure 4.10 The cloudy ring seen here indicates that the blood sample was of human origin. (From Li, R., *Forensic Biology*, Second Edition, CRC Press, Boca Raton, 2015. With permission.)

4.1.2 Semen

4.1.2.1 Components of Semen

Semen is a common body fluid encountered in cases of sexual assault. Semen is a male reproductive fluid comprised of spermatozoa and seminal fluid. Spermatozoa form in the seminiferous tubules in the testes in a process known as spermatogenesis. They are moved to the epididymis for approximately three months for maturation. After they have matured, they can be transferred from the epididymis to the ductus deferens, which is the pathway to the ejaculatory duct and urethra. The seminal fluid component of semen is a mixture of secretions from different glands. Approximately 60% of seminal fluid in semen comes from seminal vesicles, 30% from the prostate, 5% from the bulbourethral glands, and 5% from the epididymis. Seminal fluid has a normal pH between 7.2 and 7.8. Spermatozoa, the male reproductive cells, have a specific morphology which is used to aid in their identification. They are composed of the head, midpiece, and tail. The head contains the nucleus, and therefore the male DNA. The acrosomal cap on the head of the spermatozoa contains the enzymes to break into the egg for fertilization. The midpiece contains mitochondria which produce adenosine triphosphate (ATP) to energize the sperm tail, which contracts in a whip-like fashion to propel the sperm cell along the vaginal tract.

Vasectomized males still produce ejaculate, but no spermatozoa are expected to be present. In a vasectomized male, ejaculate is comprised only of seminal fluid, prostatic fluid, and bulbourethral fluid. This absence of sperm

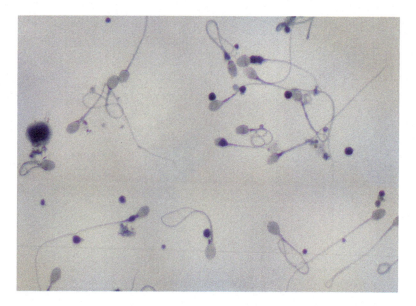

Figure 4.11 Sperm cells under the microscope. (Source: Shutterstock.com.)

cells in semen is known as azoospermia. Aspermia is a condition where no sperm cells or seminal fluid is produced. Oligospermia is an abnormally low sperm count in the ejaculate (<15–20 million sperm/mL). Reduced levels or absence of sperm cells could also result from prostate surgery, infection, genetic defects, poor nutrition, or alcohol or drug abuse. In these instances, male DNA may be found in the epithelial cells present in the seminal fluid which deposited as it traveled through the male reproductive system, but the amount of DNA present and the downstream testing processes will be different in cases with little or no sperm cells than in cases with copious sperm cells present. The presence or absence of sperm cells is vital knowledge that will determine the type of DNA extraction and amplification to be performed based on the sample type and amount of human and male specific DNA detected.

Sperm cells with tails can live inside the vagina up to four to six hours after intercourse, and sperm cells without tails can be present for up to six days in the vaginal cavity. Based on a surveillance of the literature, it is approximated that sperm cells may persist in the mouth up to six hours, in the rectum up to 20 hours, and in the cervix up to 17 days. Different proteins in seminal fluid, such as acid phosphatase, will generally degrade faster than sperm cells in the body. The length of persistence of the different components in semen will vary greatly based on the victim's activities: if they are showering, douching, eating, drinking, urinating, defecating, on their menstrual cycle, brushing their teeth, or moving around, naturally the persistence of semen will decrease. Moreover, if the victim is deceased, semen will generally be preserved longer than that in a living victim, as there is no movement, cleaning, or natural body functions, in addition to the body temperature lowering. In any cases of sexual assault, the victim's body should be tested for the presence of sperm cells and seminal fluid. A typical male will release 2.5 to 6 milliliters of semen with an ejaculation, each milliliter containing upwards of 100 million sperm. This is a rich source of DNA, which can lead to the identification of the assailant.

4.1.2.2 Presumptive Tests for Semen

Presumptive tests for semen include visual examinations, colorimetric assays, and fluorometric assays. At times, semen stains can be visible with the naked eye as a white or yellow crusty stain, or it can be discovered tactilely as it exhibits a crusty or rough texture. If semen is not immediately visible, a light source set to 420-470nm can be projected over the suspect area on an item in question. If semen is present, flavins and choline—chemicals found in seminal vesicle fluid—will fluoresce when exposed to light in that range. This fluorescence is observed by an analyst through a yellow or orange filter.

Light in this wavelength can also be used to detect saliva, sweat, and urine, so the visualization of a fluorescent spot on an object is not confirmatory for semen. Other non-biological materials may also fluoresce, such

90   Guide to the ABC Biology Exam

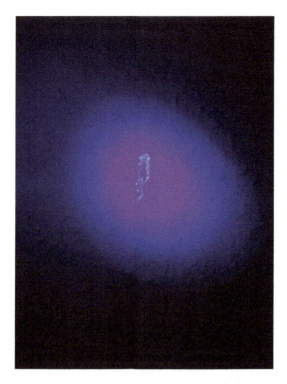

Figure 4.12 Semen fluoresces when exposed to blue light. (Image retrieved from Wikimedia, used per Creative Commons 3.0, user Utku Tanrıvere.)

as bleach, detergent, deodorant, various fibers, tooth whiteners, or certain vitamins and minerals. Colorimetric, immunochromatographic, and microscopic tests are used to further characterize a stain once detected.

There are also microscopic crystal tests for the presence of semen based on the presence of choline and spermine. In the Florence test, a suspected semen stain is extracted and placed on a slide. Florence solution (8% w/v of iodine in water containing 5% w/v of potassium iodide) is added and dark brown rosettes of rhombic crystals will form in the presence of choline. In Barberio's test, a saturated aqueous or alcoholic solution of picric acid produces yellow needle shaped crystals of spermine picrate when added to dried extract prepared on a slide. This reaction depends upon the presence of spermine from prostatic secretions.

Acid phosphatase (AP) is an enzyme produced in the prostate that is found in high concentrations in seminal fluid. The colorimetric test involves adding a drop of a solution contiaing sodium alpha naphthyl phosphate and Fast Blue B to a cutting or swab of a suspected semen stain. A purple color change within a few minutes indicates the presence of acid phosphatase. This is also known as the Brentamine test. This test can be used with a mapping technique, where an item is divided into sections. Each section is

Biological Screening

either swabbed to collect any seminal fluid present, or moist filter paper is pressed onto each section. Those swabs or filter paper are then tested for acid phosphatase. Fluorometric tests for AP are more sensitive and are useful for semen stain mapping. Moist filter paper is rubbed onto the suspected stain and then taken to a dark room and examined under ultraviolet light to detect fluorescence. The paper is sprayed with 4-methylumbelliferone phosphate (MUP) and the AP reacts immediately with blue fluorescence.

Acid phosphatase is found in other biological and non-biological material other than seminal fluid, including vaginal secretions, saliva, urine, feces, breast milk, blood, pus, nasal mucus, and certain fruits, vegetables, and fungi. Because of this lack of specificity, acid phosphatase is a presumptive test, and samples that test positive for acid phosphatase should continue on to confirmatory testing.

4.1.2.3 Confirmatory Tests for Semen

The most common way to confirm the presence of semen involves microscopic examination for the presence of spermatozoa. Cells from the suspected semen stain are transferred to a microscope slide and stained for contrast, most commonly with Christmas tree stain or hematoxylin and eosin. Christmas tree stain is composed of Nuclear Fast Red (NFR), which dyes the nuclei and acrosomal cap a pink-red color, and picroindigocarmine (PIC), which dyes the neck and tail blue-green. Epithelial cells' cytoplasm will also stain green while the nuclei will absorb the NFR taking on a red color. Hematoxylin- and eosin-stained slides will show sperm heads in purple and other cells in pink.

A newer method for separating sperm cells from non-sperm cells is through the use of a laser capture microdissection (LCM) machine. This is beneficial when dealing with mixed samples from a female victim and male perpetrator in sexual assault cases. A thin polymer is placed on an LCM cap, which is then placed over fluorescently detected spermatozoa cells on a microscope slide made of polymer. A laser melts the polymer, and it adheres only to the spermatozoa, which can then be lifted off the slide and placed into tubes for DNA analysis. This technology is useful because it can be fully automated, freeing up live analysts for other examinations.

Another widely accepted method for confirming the presence of seminal fluid is testing for the presence of prostate-specific antigen (PSA), also known as p30. PSA is a protein produced in the prostate that is found in high concentrations in seminal fluid. Small quantities of p30 can also be found in other body fluids, such as male urine and blood, breast milk, vaginal secretions, and amniotic fluid.

Seminal vesicle-specific antigen (SVSA) is a protein that coagulates semen into a gel-like substance on ejaculation and can also be used to identify semen. SVSA consists of two major types: semenogelin I (SgI) and

Figure 4.13 Christmas tree stain is commonly used when searching for sperm on a microscope slide. (Used courtesy, and with permission of, executive director and chief forensic serologist, Serological Research Institute, Richmond, California.)

Figure 4.14 A positive PSA immunochromatographic test. (Source: Shutterstock.com.)

semenogelin II (SgII). Semenogelin is present in body tissues such as skeletal muscle, kidney, colon, and trachea, but the only forensically significant fluid it is found in is seminal fluid. It can be detected using Monoclonal Mouse anti-Human-Sperm antibody number 5 (MHS-5). Confirmatory tests for the presence of PSA and SVSA are most commonly immunochromatographic

Biological Screening

cards; however, they can also be detected by immunodiffusion, immunoelectrophoresis, RT-PCR, and Enzyme Linked Immunosorbent Assays (ELISA tests). Similar to blood confirmation, high-dose hook effect can also affect immunochromatographic assays for semen. Commercial immunochromatographic assays available for semen include the PSA-check-1, Seratec® PSA Semiquant, and One-Step ABAcard PSA®.

4.1.3 Saliva

Saliva is a water-based substance containing enzymes, electrolytes, glycoproteins, and antibodies present in the mouth and human digestive tract. Human saliva has a pH of 7.4 in a healthy person. Around 1.0–1.5 liters of saliva are produced daily in the human salivary glands, mainly during eating. Saliva is a fluid that may be left behind in cases of sexual assault. To detect saliva, visual examinations are first performed followed by colorimetric assays or starch-iodine assays. If it is not visible to the naked eye, saliva will fluoresce under a wavelength of 420-470nm through a yellow or orange filter. It can also be examined microscopically following staining to identify buccal epithelial cells.

Amylases are the enzymes in saliva that break down D-glucose polymers at the α1-4 linkage sites on carbohydrates. Two types of amylases can be characterized: β-amylases and human α-amylases. β-amylases are found in plant and bacterial sources and therefore their presence is not indicative of human saliva. Human α-amylases, on the other hand, are found in two major isoenzyme forms: human salivary α-amylase (HSA) and human pancreatic α-amylase (HPA). The detection of amylase in a sample is a cornerstone in saliva identification. Since all forms of amylase are largely homologous, presumptive tests indicate the presence of amylase but cannot distinguish HSA from HPA or α- from β-amylases. Confirmatory tests involve detection of HSA proteins through immunochromatographic cards, ELISA, and RT-PCR.

4.1.3.1 Presumptive Tests for Saliva

Colorimetric assays are conducted by adding suspected saliva to a solution of amylose or amylopectin. The HSA in the saliva breaks down the amylose or amylopectin and the solution changes color in proportion to the amount of amylase in the sample. A commonly used substrate for presumptive saliva identification is the Phadebas reagent. This can be used in one of two ways. First, a Phadebas tablet can be added to a tube with a small cutting of the presumed saliva stain. The tube is incubated and the pH is increased by adding sodium hydroxide. Inside the tablet, there is a blue dye that is conjugated to starch. In the presence of amylase, the starch is digested, releasing the blue dye. The blue color can be measured at 620nm using a spectrophotometer and the measurement of optical density can be converted to amylase units

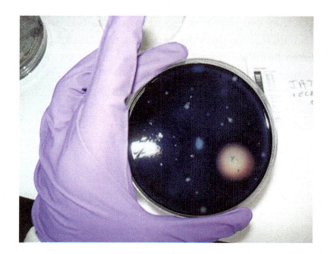

Figure 4.15 Samples are introduced to an agarose gel plate, and if saliva is present the starch metabolizes the iodine in the gel. (Used courtesy, and with permission of, Stephen Gallagher, Harry S. Truman High School.)

by comparison to a standard curve. A second application of the Phadebas reagent is to spray it onto a sheet of filter paper, then press the filter paper onto an item suspected of bearing saliva. Plastic wrap and a heavy object are placed on top of the paper during development and the paper is observed every minute for 10 minutes, then every 5 minutes until 40 minutes are reached. A positive reaction is indicated by a light-blue area on the paper within these 40 minutes. SALIgAE® (Abacus Diagnostics) is another common colorimetric assay used to detect saliva. It is also available in a spray kit for amylase mapping.

Starch-iodine assays involve injecting suspected saliva into a well in an agarose gel plate. The saliva is allowed to diffuse through a starch-based gel. The gel is then stained by adding iodine and the color is observed. If saliva is present, the amylase will break down the starch around the sample well leaving a clear "zone of digestion." The size of the clear zone is proportional to the amylase quantity in the sample.

4.1.3.2 Confirmatory Tests for Saliva

Confirmatory tests for saliva include immunochromatographic assays, ELISA, and RT-PCR. The RSID™-Saliva kit (Independent Forensics) is a commercial immunochromatographic kit where a labeled monoclonal anti-HSA antibody is present in the sample and test wells to detect HSA. Like other immunochromatographic assays, saliva testing can also be affected by high-dose hook effect. More than 50 microliters of neat saliva will produce a high-dose hook effect, or a false negative result. ELISA tests for saliva utilize the antibody-antigen-antibody sandwich as well, where a positive result will

Biological Screening

Table 4.1 Immunochromatographic Cards Utilized in Forensic Testing

Assay	Antigen	Labeled Antibody	Immobilized Antibody	Forensic Application
ABAcard® HemaTrace® (Abacus Diagnostics)	Hemoglobin (Hb)	Monoclonal antihuman Hb antibody	Monoclonal or polyclonal antihuman Hb antibody	Blood and species identification
RSID™-Blood (Independent Forensics)	Glycophorin A (GPA)	Monoclonal antihuman GPA antibody	Monoclonal antihuman GPA antibody	Blood and species identification
RSID™-Saliva (Independent Forensics)	Human salivary α-amylase (HSA)	Monoclonal antihuman HSA antibody	Monoclonal antihuman HSA antibody[a]	Saliva identification
One-Step ABAcard PSA® (Abacus Diagnostics)	Prostate-specific antigen (PSA)	Monoclonal antihuman PSA antibody	Monoclonal or polyclonal antihuman PSA antibody	Semen identification
RSID™-Semen (Independent Forensics)	Semenogelin (Sg)	Monoclonal antihuman Sg antibody	Monoclonal antihuman Sg antibody[a]	Semen identification

From Li, R., *Forensic Biology* (2nd ed.). CRC Press, Boca Raton, 2015. With permission.

[a] The epitope recognized by the immobilized antibody is different from that of the labeled antibody.

produce either a colorimetric or a fluorometric signal whose intensity corresponds to the amount of bound antigen. RNA-based assays utilizing RT-PCR have recently been developed for saliva identification by identifying mRNA solely expressed in cells in the oral cavity.

4.1.4 Urine

4.1.4.1 Properties of Urine

Urine is a byproduct of blood filtered through the kidneys. It primarily consists of water, salts, glucose, and urea. A normal pH for urine is about 6.0. Urine is formed in the nephrons, which are found in the kidneys. Blood is filtered through the glomeruli of the nephron, where its fluid flows through the capillaries into the Bowman's capsule. The filtered fluid that comes from the Bowman's capsule enters the renal tubule for reabsorption, where much of the water, glucose, nutrients, and ions re-enter the blood. The remaining fluid containing ions, ammonia, and metabolites is drained from the kidney through the ureter and stored in the bladder until its release through the urethra. Urine may be found in sexual assault, harassment, and breaking and entering cases. Urine stains found at homicide scenes can suggest information as to the manner of death; death by ligature or manual strangulation typically causes victims to involuntarily urinate prior to death. Though urine is typically a poor fluid for developing a DNA profile, the detection of urine can be helpful for reconstructing crime scenes and confirming witness accounts.

4.1.4.2 Presumptive Tests for Urine

Urine can be characterized by the yellow color and the unique smell. As with many forensic analyses, the analysts' first tools of examination are their eyes and nose. If the naked eye fails to detect it, urine will fluoresce under UV or blue light. All presumptive tests for urine are based on the identification of urea, creatinine, and uric acid.

DMAC and colorimetric assays are the most common presumptive tests for urine. DMAC is short for p-dimethylaminocinnamaldehyde, a chemical that will turn pink within 30 minutes of interaction with the urea in a urine sample. DMAC can also be used fluoromctrically, where a filter paper is dampened with DMAC and applied to the stain, wrapped in aluminum foil, and allowed to sit overnight. If urine is present, the sample will fluoresce at 473–548nm. Unfortunately, DMAC reacts positively with saliva, semen, sweat, and vaginal secretions as well. Diluted DMAC will react only with urine due to the high urea content not present in the other body fluids, so dilution can increase specificity. Urease assays are colorimetric and detect urea by use of an acid-base indicator known as bromothymol blue, which turns blue upon reaction with urea. Urea can also be detected with manganese and silver nitrates, which produce a black color when a positive reaction occurs.

Identification of creatinine is another method for the presumptive testing of urine. Creatinine is a byproduct of metabolism that is released from the muscles into the blood. Creatinine concentrations in urine are proportional to the muscle mass of the individual. However, it is also present in blood and semen. Testing for creatinine is performed with the Jaffe color test, which involves the creation of creatinine picrate, a bright-red substance. A commercial device known as Uritrace (Abacus Diagnostics) is also available for creatinine detection.

4.1.4.3 Confirmatory Tests for Urine

Confirmatory assays for urine include identification of Tamm-Horsfall protein (THP) and 17-ketosteroids. THP, or uromodulin, is the protein most prevalent in urine. Typically, adults will excrete 20–100mg of THP each day. THP has only been identified in urine, thus a positive identification of THP is a confirmation for urine. THP is tested for by ELISA or immunochromatographic assays such as RSID™-Urine. 17-ketosteroids exist in urine as the metabolite form, as they undergo glucuronidation and sulfation in the liver. Five forms of 17-ketosteroids are identified using liquid

Biological Screening 97

chromatography—mass spectrometry (LC—MS) to confirm the presence of urine. 17-ketosteroid profiles are distinguishable between humans and animals.

4.1.5 Feces

Feces are the waste material produced from digested food. They consist of bacteria, undigested materials, sloughed intestinal epithelial cells, electrolytes, bile pigments, and water. Feces are formed in the intestines from the remainder of whatever substances are not absorbed by the intestines. Feces found at a crime scene can indicate sexual assault involving sodomy, assault with fecal matter, or vandalism. Feces can be detected by visual and olfactory examination, by use of the urobilinoids test, or by identifying the bacteria in the sample (typically Bacteroides). Color and odor of feces are primary characteristics for identification. Microscopic examination of feces will often show vegetable and meat fibers.

The Edelman and Schlesinger tests are used to detect urobilinoids, which are proteins created in the degradation of heme that produce the brown color in fecal matter. Both the Edelman and Schlesinger tests are fluorometric, producing a bright "apple green" color in the presence of urobilinoids. Neither test can distinguish between human and other mammal feces and both can cross react with urobilin in urine stains. Bacteria identification in feces looks for Bacteroides through the use of RT-PCR; however, it cannot distinguish human and animal feces. Once a sample is identified as feces, it is possible the sloughed intestinal cells can be used to extract a DNA profile.

Bibliography

James, S. H., Nordby, J. J., & Bell, S. (2014). *Forensic science: an introduction to scientific and investigative techniques* (4th ed.). Boca Raton: CRC Press.

Li, R. C. (2015). *Forensic biology* (2nd ed.) Boca Raton: CRC Press (Ch 11, 14, 16, 17).

Saferstein, R. (2018). *Criminalistics: an introduction to forensic science* (12th ed.). Boston: Pearson (Ch 5, 14).

Anatomy and Cell Biology

5

5.1 Anatomy, Physiology, Reproductive Biology

5.1.1 Biochemistry of Physiological Fluids

In Chapter 4, we discussed the components and biochemistry of forensically significant physiological fluids. The delicate biochemical makeup of each fluid is vital to its success and there is a great deal that can be learned from slight changes in each fluid. For example, normal seminal fluid pH is between 7.2 and 7.8. The pH of semen is slightly more basic than that of water, which has a pH of seven. The slightly basic pH of semen is designed to increase the chances of reproductive success by protecting the spermatozoa present in the ejaculate from the highly acidic environment in the female reproductive tract. Acidic ejaculate (lower pH value) may indicate one or both of the seminal vesicles are blocked. A basic ejaculate (higher pH value) may indicate an infection. The specific pH, cellular composition, and morphological design play important roles in the success or failure of the fluid to function as intended.

5.2 Cellular and Molecular Biology

5.2.1 Cell Morphology

Most somatic cells have a basic structure. They have a cell membrane containing cytoplasm, organelles, and a nucleus. A common misconception among lay people is that red blood cells contain DNA. Since mature red blood cells are anucleated, they contain no DNA. The leukocytes, or white blood cells, are the source of the DNA that comes from a bloodstain. The nucleus is considered the central and most important part of the cell. If we think of an egg as being analogous to a cell, the nucleus is the yolk. The DNA, or the blueprint for creating all the proteins vital for the cell's survival, is located in the nucleus.

There are two main types of cells: prokaryotic and eukaryotic. Prokaryotic cells are single-celled organisms that lack any defined organelles or nuclei. Examples of prokaryotic cells are bacteria, which are beginning to take on a more significant forensic role. Prokaryotic bacteria cells have double-stranded, circular DNA rather than the chromosomal DNA that exists in the eukaryotic

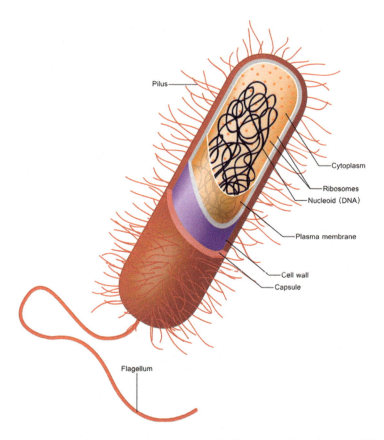

Figure 5.1 Typical prokaryotic bacteria cell. (Image retrieved from Wikimedia, used per Creative Commons 4.0, user Ali Zifan.)

cells. Prokaryotic cells are much simpler than eukaryotic cells, and much forensic focus will be on the eukaryotic cells of the human body. Eukaryotic cells have several physical structures that differentiate them from prokaryotic cells. They have a cell membrane that contains the cell contents, a nucleus that holds the DNA, and organelles that maintain various aspects of cell function. With the advent of nuclear DNA and the advancements in technology made over the last three decades, forensic scientists need only a few of these cells to develop a DNA profile.

Within the delineation of eukaryotic cells, we also find plant and animal varieties. In forensic molecular biology, we are examining human body fluids and tissues that contain eukaryotic animal cells, but there are also areas of forensic science that concentrate on plants. DNA from plants can be extracted and compared just like human biological DNA. The major differences between plant and animal eukaryotic cells include: the cell wall, which tends to be more rigid when compared to the cell membrane of a human somatic cell, and the presence of chloroplasts, which are organelles absent in animal cells.

Anatomy and Biology

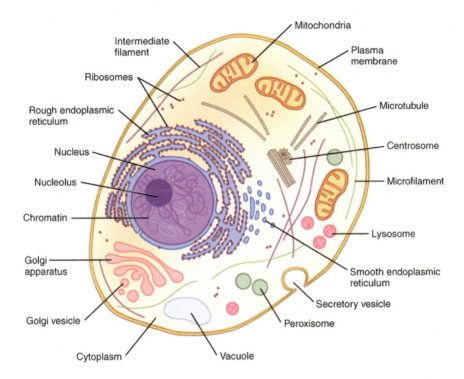

Figure 5.2 Structure of an animal cell showing many basic components, including the nucleus, organelles, and plasma membrane. (Image retrieved from Wikimedia, used per Creative Commons 4.0, user OpenStax.)

In cases where regular nuclear DNA methods are unsuccessful, the mitochondrion becomes a forensically significant organelle. Most somatic body cells contain mitochondria, which are known as the "powerhouses" of the cell. They are responsible for respiration and energy production within the cell. It is estimated that there are between 1,000 and 2,000 mitochondria in each cell, and each mitochondrion has its own circular, double-stranded DNA. The sheer number of mitochondria, each containing its own DNA, means there is a high volume of mitochondrial DNA present in the body. In cases of degraded DNA, skeletal remains, and hair evidence, mitochondrial DNA testing may be the last and final hope of obtaining genetic information.

Cell morphology refers to the shape and appearance of cells. Cells can vary largely in size, components, and function. Some cells require things like oxygen, water, or sunlight. Each component of the cell is specifically designed to perform the exact functions that are required for that space, fluid, or organism. Forensically, the morphology of certain types of cells is vitally important. An easy example of this is the sperm cell. The morphological structure of the cell itself helps forensic biologists to identify it from other cells. Analysts can rely on the presence of a clearly defined acrosomal cap

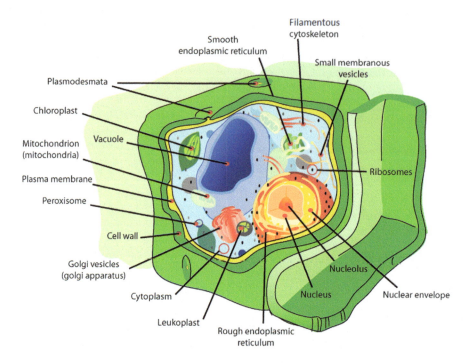

Figure 5.3 Eukaryotic plant cell structure.

Figure 5.4 Buccal epithelial cells are commonly encountered cells in forensic science; note the nucleus stained darker purple in the center. (Image retrieved from Wikimedia, used per Creative Commons 4.0, user Dr. Shikha Jaggi.)

Anatomy and Biology

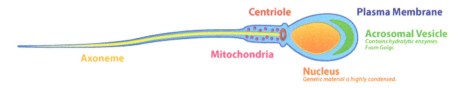

Figure 5.5 Sperm cells are the male reproductive cells that use their long flagellum to propel themselves to the egg. (Image retrieved from Wikimedia, used per Creative Commons 3.0, user Studentreader.)

to differentiate sperm cells from other similarly sized and shaped cells that might be present in the vaginal cavity, such as yeast cells. Forensic biologists also exploit the structure of the sperm cell membrane during the differential extraction. The sperm cell needs a much tougher cell membrane in order to protect it from the environmental insults it will face in the female reproductive tract. The disulfide bonds in the tough sperm cell membranes are much harder to break than the plasma membranes in epithelial cells. This knowledge allows for scientists to apply two different chemicals of increasing strength, first lysing the weaker cells and separating the lysate from any intact sperm before lysing those.

5.2.2 Cells and Chromosomes

Eukaryotic cells contain a rich network of organelles, all of which serve complex and vital purposes in cell function. DNA has become a main focus in forensic science over the last 20 or 30 years. DNA is present in prokaryotic cells in a single loop; tightly wound in chromosomes in the nucleus of eukaryotic cells; or in the mitochondria of eukaryotic cells. Nuclear DNA is the preferred and most discriminating source of forensic DNA and is wound into chromosomes. Chromosomes are organized in pairs, one half of the pair inherited from the mother and one half from the father.

Chromosomes have a short arm (p, for petit) and a long arm (q, for queue). The arms are connected by a centromere. Chromosome ends are known as telomeres. Chromosomes contain DNA wound around histone proteins and non-histone chromosomal proteins. The DNA in each chromosome exists as a single, unbroken strand. It is wrapped around the histones to compact it into nucleosomes, which are then compacted further into chromatin.

104   Guide to the ABC Biology Exam

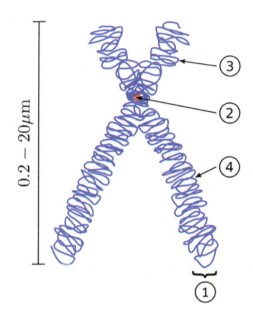

Figure 5.6 1) Chromatid 2) Centromere 3) Short arm 4) Long arm. (Image retrieved from Wikimedia, used per Creative Commons 3.0, user Original version: Magnus Manske, revised version user Dietzel65.)

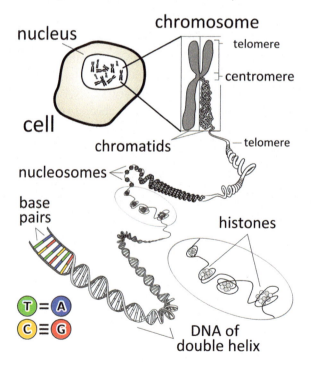

Figure 5.7 DNA is wound around histone proteins to form nucleosomes, which wind around each other to form chromosomes.

Anatomy and Biology

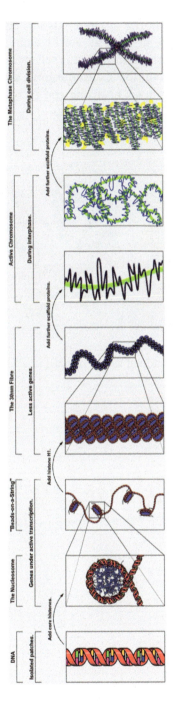

Figure 5.8 A second depiction of DNA being wound into chromosomal form.

Chromatin can be found in two forms: euchromatin and heterochromatin. Euchromatic areas are condensed and released regularly during the cell reproduction cycle. The majority of DNA exists as euchromatin, and this is the DNA that contains the genes that can be expressed. Heterochromatic regions typically stay condensed through the cell cycle and can be found near centromeres, the long arm on the Y chromosome, and in the short arms of chromosomes with centromeres near one end (acrocentric chromosomes). Heterochromatic genes are typically inactive.

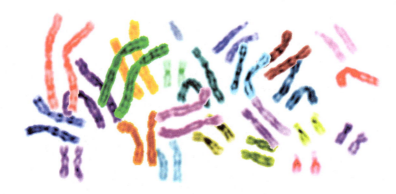

Figure 5.9 Chromosomes stained to be arranged in a karyotype.

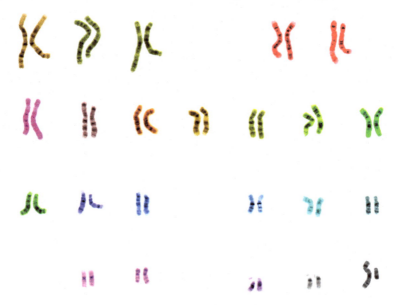

Figure 5.10 Stained chromosomes are separated and organized with their sister chromatid in descending size order (with the exception of chromosome 21 being smaller than chromosome 22).

Anatomy and Biology

5.2.3 Chromosomal Organization

Human diploid cells contain 22 pairs of chromosomes and 2 sex chromosomes, 46 chromosomes in total. Twenty-three chromosomes come from the mother, and 23 from the father. The mother provides an X chromosome in the ovum, and the father provides either an X or a Y chromosome in the spermatozoon to determine the sex of the baby (XX for female, XY for male). Chromosomes can be stained and organized into a set from largest to smallest, called a **karyotype**. In a karyotype, chromosomes are given a number based on their size, number 1 being the largest, with an exception: chromosome 21 is smaller than 22. Karyotypes not only show the sex of the DNA donor, but can also show any abnormalities in the chromosome(s) on which they exist. For example, Down syndrome is caused by the presence of an extra chromosome number 21.

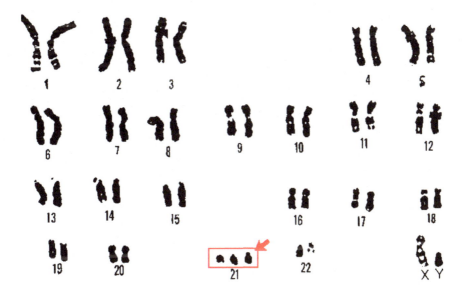

Figure 5.11 Karyotype of an individual with trisomy 21, causing Down syndrome.

5.2.4 Cellular DNA Content

Each chromosome contains thousands of genes, which are composed of a series of nucleotides. There are four nitrogenous bases that comprise human DNA: adenine (A), thymine (T), cytosine (C), and guanine (G). Adenine pairs with thymine, forming a double hydrogen bond between them. Cytosine pairs with guanine, forming a triple hydrogen bond. The DNA base pair sequences found on these chromosomes are unique to each individual.

Eukaryotic chromosomes are found in the nucleus of the cell, and all 46 chromosomes are present in all cells of the body except for the reproductive cells, which contain only 23 chromosomes. DNA not only provides the blueprint for physical and mental characteristics, but it also codes for proteins that regulate and control all of the functions in the body. Eukaryotic cells have DNA present in chromosomal form in the nucleus, as well as in the mitochondria. Mitochondrial DNA is passed only from mother to child in its entirety. Every one of the thousands of mitochondria in the cell has a complete copy of the mitochondrial DNA.

5.2.5 Cell Division

All cells are born from the process of cell division, known as **mitosis**. The cell cycle contains four main phases: cell growth (G1 phase), chromosome duplication (S phase), continued cell growth (G2 phase), and cell division (M phase). Mitosis is the creation of two identical daughter diploid cells from one parent diploid cell. The division of gametes (egg and sperm) is known as meiosis, which will be discussed later.

This cell cycle varies in time from organism to organism. The first three phases are known as interphase, which takes up the majority of the time of the cycle. The final phase of the meiosis process consists of mitosis and cytokinesis, and typically takes about one hour in mammals. Mitosis is divided into four stages: prophase, metaphase, anaphase, and telophase. In prophase, chromosomes align with their replicate chromosome, and the nuclear envelope begins to dissolve. Mitotic spindles form out of the two centrosomes, which move to opposite poles of the cell. In metaphase, the chromosomes move to the equator of the cell between the centrosomes and the mitotic spindles attach to the kinetochore of the chromosomes. In anaphase, the chromosomes are separated and are pulled to opposite ends of the cell as the

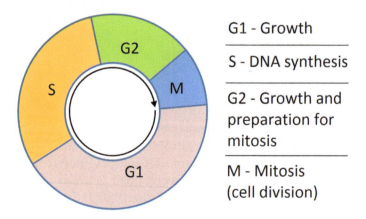

Figure 5.12 The cell cycle, showing the relative amount of time it takes for each stage of the cycle.

Anatomy and Biology

mitotic spindles shorten. Finally, in telophase, the chromosomes reach the poles of the cell, nuclear envelopes form, and the cell begins to pinch inwards in the middle. Cytokinesis is the process by which the cytoplasm is divided in two by the efforts of actin and myosin filaments, thus creating two identical daughter cells.

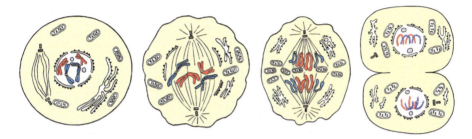

Figure 5.13 The phases of mitosis: prophase (far left), metaphase (center left), anaphase (center right), telophase (far right).

5.2.6 DNA Structure

The double helix structure of DNA was first described by James Watson and Francis Crick in the 1950s. The two scientists based their discoveries of the structure of DNA on X-ray diffraction images produced by Rosalind Franklin, an expert in X-ray diffraction working at King's College in London. The helix is composed of two complementary chains of nucleotides held together by hydrogen bonds. These chains are composed of repeating units known as nucleotides. Nucleotides consist of a nitrogen-containing base, a five-carbon sugar, and a phosphate group. The sugars are deoxyribose sugars, and the bases can be one of four variations: adenine, thymine, cytosine, and guanine.

The DNA structure has the appearance of a ladder that has been grabbed by the legs and twisted, with the two edges consist of alternating sugars and phosphate groups, and the "rungs" consist of the bases—adenine attaching to thymine, and cytosine attaching to guanine, each held together by hydrogen bonds. The G-C bonds are triple hydrogen bonds and the A-T bonds are double hydrogen bonds. For this reason, G-C-rich stretches of DNA have a higher melting point than A-T-rich sections.

There are three major structural forms of DNA: A-DNA, B-DNA, and Z-DNA. The B form, described by Watson and Crick, is the most common and is believed to predominate in cells. It has a right-hand twist with approximately 10.5 base pairs making up each turn. A-DNA also has a right-hand helical twist and the turns contain approximately 11 DNA base pairs. Z-DNA has helical strands that turn the opposite way to A-DNA and B-DNA. Z-DNA has a left-handed helical twist, and approximately 12 base pairs make up each turn.

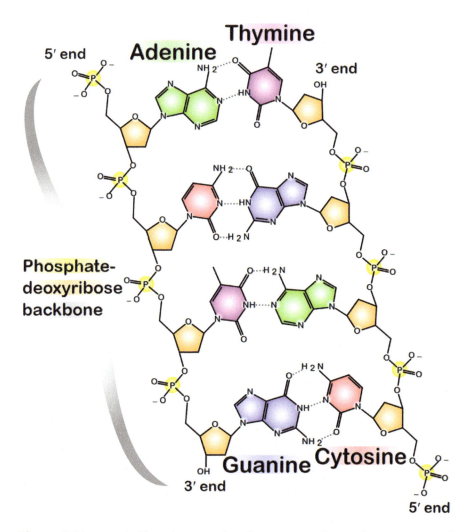

Figure 5.14 DNA ladder, showing phosphate groups, deoxyribose sugars, and nitrogen-containing bases.

In B-DNA, there are major grooves and minor grooves in the helix. Twin helical strands form the DNA backbone. Another double helix may be found by tracing the spaces, or grooves, between the strands. These voids may provide binding sites for DNA replication. Because the strands are not directly opposite each other, the grooves are unequally sized. The narrowness of the minor groove means that the edges of the bases are less accessible to the DNA transcription machinery. The bases are more accessible in the major groove. From a practical standpoint, proteins such as transcription factors that can bind to specific sequences in double-stranded DNA usually make contacts to the sides of the bases exposed in the major groove. The DNA amplification machinery used for forensic purposes must also be able to access the bases

Anatomy and Biology 111

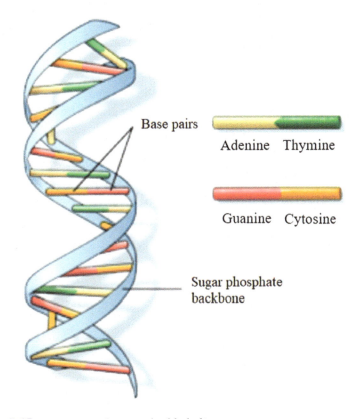

Figure 5.15 DNA wound into a double helix.

at each selected location. Each location on the DNA that has been selected for forensic purposes has been studied, and accessibility of the location is a deciding factor in whether or not it should be widely used.

5.2.7 Transcription and Translation

Transcription and translation are the steps that transform DNA to RNA, and RNA to protein. This process demonstrates how the DNA sequence is used as a blueprint for making all the things the body needs to function properly. Transcription always occurs first, transforming DNA to RNA. The structure of DNA is different from RNA in that RNA is only single-stranded, unlike DNA's double-stranded form. RNA also contains a ribose sugar rather than a deoxyribose sugar, and uracil instead of thymine. Transcription occurs by opening and unwinding the DNA strand of interest and creating a complementary RNA strand using one side of the DNA as a template. RNA polymerase is responsible for opening the DNA strands by breaking the hydrogen bonds between nucleotides, as well as linking the ribonucleotides together covalently to form the RNA strand. Next, the RNA is capped on one end with specific nucleotides, given a "tail" on the other end with multiple adenosine nucleotides, and spliced to form mature RNA. Once the RNA strand is complete, it is known as messenger RNA (mRNA), as it carries the "message" or code for the protein that is to be made.

From here, the mRNA travels out of the nucleus to the cytosol, where it attaches to a ribosome, and translation begins. The ribosome reads the mRNA strand three nucleotides at a time, and transfer RNA (tRNA) attaches the amino acid coded for by those three nucleotides. Once the entire mRNA strand has been read by the ribosome, the amino acids fold together to create the protein. While this process can easily be summarized, there are countless components that play a role in the process of transforming DNA to protein, many of which are still unknown to scientists.

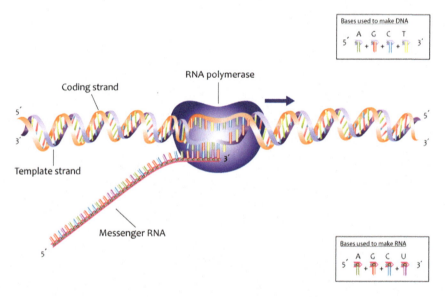

Figure 5.16 Transcription. The DNA is processed by the RNA polymerase, creating an mRNA strand complementary to the DNA.

Anatomy and Biology

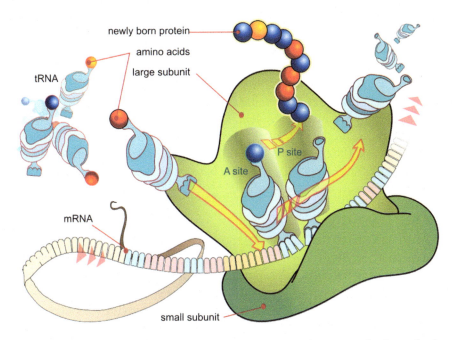

Figure 5.17 Translation. The mRNA attaches to a ribosome, which reads the strand and creates a protein based on the nucleotide sequence.

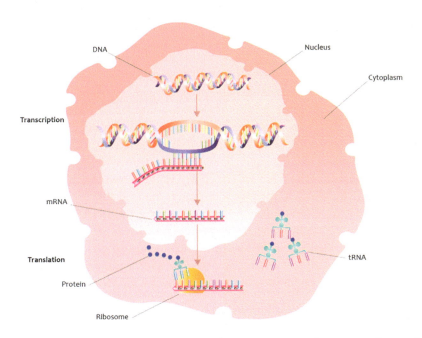

Figure 5.18 The transcription and translation processes shown together in the same cell.

5.2.8 Replication

When DNA replicates, it uses the complementary base pairing inherent to its structure for guidance. DNA is made up of two strands, which are linked together by hydrogen bonds and twisted in the helical shape. When DNA replicates, it splits apart. Each separate strand of the DNA acts as a template for the other to produce a complementary strand. Every time a cell divides, it must replicate all the DNA within. In context, the process takes place over eight hours and the equivalent of 1,000 textbooks are copied, with only a few letters out of place. Approximately 100 base pairs are duplicated per second in human replication with exceptional speed and accuracy.

5.2.8.1 DNA Organization

In order to unwind the helix and expose the unpaired bases, the replication origins must be located by the initiator proteins. These break the hydrogen bonds, exposing the bases to be copied. The human genome has approximately 10,000 replication origins because the genome is so long. Each chromosome has around 220 replication origins to facilitate faster copying. Once the initiator protein binds to the replication origin, it attracts the replication machinery.

Chromosomes during interphase are longer than chromosomes during mitosis as the chromosomes wind up more tightly during the process of mitosis. During interphase, the chromosomes are organized in such a way that they do not become entangled with one another too extensively.

5.2.8.2 Replication Forks and Bubbles

When DNA is being copied, it takes on a Y shape called a "replication fork." Two replication forks are formed at each origin. It is at these forks that the DNA is unzipped and each side is being "rezipped" to a new piece of DNA being synthesized. The new strand of DNA is compiled by DNA polymerase.

Using one side of the original DNA strand as a template, the polymerase adds the base pairs that complement to make a whole new complementary strand. Each strand will undergo the replication process with its own replication fork. One side of the DNA will be copied in the 3' to 5' direction, while the other side will be copied in the 5' to 3' direction. Because the same polymerases complete the replication process in both directions, the DNA strand in the 3' to 5' direction is actually copied in pieces (called Okazaki fragments), and the pieces are assembled into a full strand. The strand being copied and linked simultaneously in the 5' to 3' direction is called the "leading strand," while the strand being copied and linked discontinuously is called the "lagging strand."

Anatomy and Biology

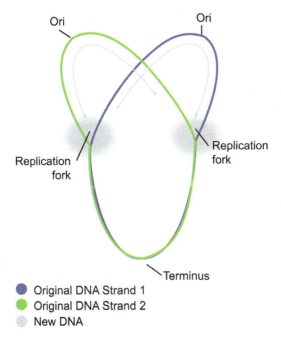

Figure 5.19 Replication of a circular strand of DNA showing two replication forks moving in opposite directions.

5.2.8.3 Enzymes Involved in DNA Replication

DNA polymerase is the primary enzyme involved in DNA replication. It moves along the unzipped DNA strand and adds the nucleotides onto the

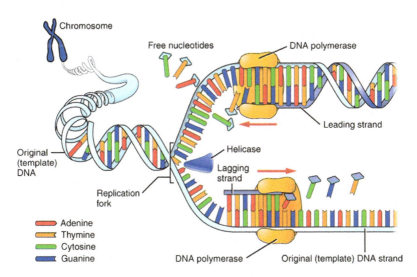

Figure 5.20 The enzymes involved in DNA replication, all at work together simultaneously.

side chains to create two daughter strands from the one parent strand. DNA polymerase always moves in the 5' to 3' direction, and it contains a proofreading mechanism that nearly eliminates all errors in replication.

DNA helicases unzip the double helix, while single-strand DNA-binding proteins stick onto the exposed DNA strand to keep them from re-attaching until the DNA polymerase reaches it. As the DNA is unwound by the DNA helicase, the helix gets wound tighter, and tension must be relieved. This is done by DNA topoisomerases. Sliding clamp enzymes keep the DNA polymerase firmly attached to the DNA strand so it doesn't fall off prematurely in the replication process. Telomerase is the enzyme that signals the end of the DNA replication.

5.2.8.4 *Proofreading Mechanisms*

DNA polymerase contains a proofreading mechanism that allows it to correct for the rare error made in DNA replication. DNA polymerase is self-correcting, which is what gives the replication process speed and accuracy. It is estimated that a replication error occurs only once in every 10^7 base pairings. Even with that low mutation rate because the outcomes of these errors can be so catastrophic, DNA polymerase proofreads the nucleotide additions during DNA synthesis to ensure the proper base pair has been added. If it finds the wrong base pair has been added in error, the base is cleaved by the enzyme and a new, complementary base is added. This proofreading function is only possible in the 5' to 3' direction, which explains why the lagging strand must first be synthesized in pieces and then linked together.

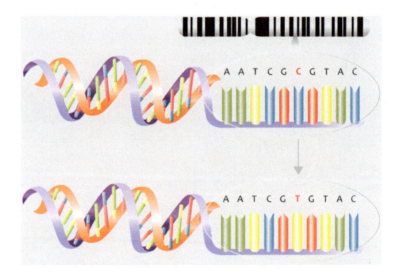

Figure 5.21 A substitution mutation, where the parent DNA was misread and a T was inserted instead of a C.

Anatomy and Biology

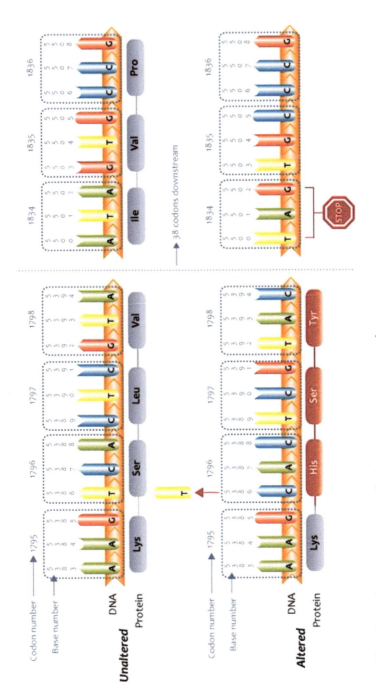

Figure 5.22 A frameshift mutation, producing a premature stop codon.

5.2.9 Mutation Mechanisms and Rates

Mutations are errors that occur in the DNA replication process and produce daughter strands that are different to the parent strands. Mutations in the DNA sequence can prove to be beneficial, hurtful, or neutral to the organism. Most commonly, mutations cause more harm than good, but occasionally the mutation may have no effect on the organism.

5.2.9.1 Kinds of Mutations

Mutations can occur in a variety of ways. The errors mainly occur during DNA replication and transcription.

5.2.9.1.1 Substitutions and Frameshifts

A substitution mutation is where one nucleotide is mistakenly entered into the DNA sequence instead of the correct nucleotide. A frameshift is when one or more nucleotides are added or subtracted between codons. The result of this is that when the DNA becomes transcribed and translated, the resulting protein is inaccurate because the codons have shifted.

5.2.9.1.2 Unstable Trinucleotide Repeats and Other Repeat Mutation Mechanisms

Often in the DNA sequence, nucleotides will repeat themselves a number of times. Unstable trinucleotide repeats occur when a set of three nucleotides naturally repeats a certain number of times, but the mutation causes it to repeat many more times. The mutation causes an unstable number of repeats that typically leads to serious genetic diseases.

5.2.9.1.3 Unequal Crossing Over

Unequal crossing over occurs during homologous recombination, where there is a misalignment between homologous chromosomes so that the resulting chromosomes are unequal in size. This causes some chromosomes to have two sets of the same gene on the same chromosome, and some chromosome to have no set or a partial set of the gene.

5.2.9.1.4 Insertions/Deletions

Insertions include one or more nucleotides being added into the DNA sequence, while deletions include one or more nucleotides being excised from the DNA sequence. If a series of three nucleotides are either inserted or deleted, this will only alter the protein produced through transcription and translation by one amino acid, since it will shift the sequence by only one codon. However, if the insertion or deletion occurs with a number of nucleotides not divisible by three, it can potentially change all subsequent amino acids following the mutation, creating a completely different protein and most often causing difficulty.

Anatomy and Biology

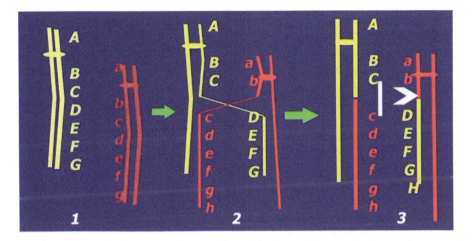

Figure 5.23 Unequal crossing over between two chromosomes, creating one chromosome with two "C" loci and one chromosome with no "C" loci.

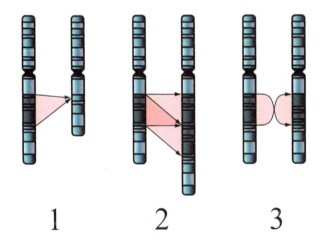

Figure 5.24 1) Deletion 2) Duplication 3) Inversion.

5.2.9.1.5 Rare Mutations
Certain mutations are known as de novo mutations, indicating that they arise spontaneously. These mutations are observed in about 1% of the population and are the source of disorders such as autism and schizophrenia. The fact that these mutations occur spontaneously is the only reason they still remain in the population, since the affected individuals rarely leave descendants.

5.2.9.2 *Repair*
To repair the damaged or mutated DNA, first it must be recognized that the DNA contains a mutation. Nucleases then cleave the covalent bonds to separate the damaged nucleotides from the DNA strand. A repair DNA

polymerase binds to the 3' end of the gap and fills in the proper nucleotide sequence. Finally, DNA ligase seals the break in the sugar-phosphate backbone that was made by the nucleases, and the mutation is successfully excised.

Double-strand mutations are much more difficult to repair. One method is to put the broken ends of DNA back together in a process called nonhomologous end joining, which repairs the damage rapidly, but often nucleotides at the repair site are lost in the process. Homologous recombination is a much more successful double-strand repair method, in which a nuclease digests the 5' ends of the broken strands and the DNA performs complementary base-pairing with the other daughter strand, then uses the complementary strands as a template to repair the DNA.

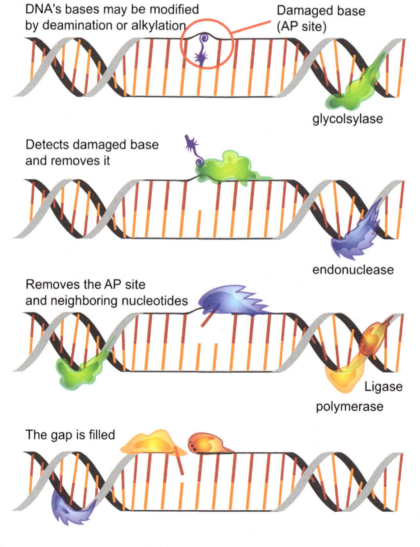

Figure 5.25 The process involved to repair a damaged base on DNA.

Anatomy and Biology

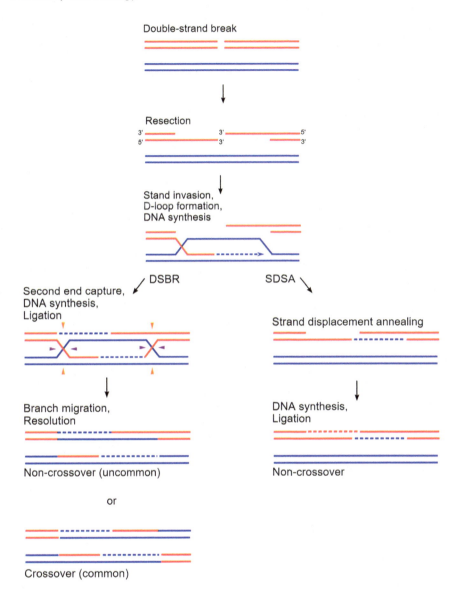

Figure 5.26 Homologous recombination to repair double-stranded DNA breakage.

Bibliography

Alberts, B., Bray, D., Hopkin, K., Johnson, A., Lewis, J., Raff, M., & Roberts, K. (2014). *Essential cell biology* (4th ed.). New York: Garland Science (Ch 1, 5, 6, 9, 18, 19).

Saferstein, R. (2018). *Criminalistics: an introduction to forensic science* (12th ed.). Boston: Pearson (Ch 10).

Concepts in Genetics, Biochemistry, and Statistics

6.1 Genetics

In this section, we will focus on the path of genes through sexual reproduction, on some unique genes in cells, and on problems that may arise in the process of gene transfer. There are two types of nucleated cells in the human body: haploid and diploid. The main difference between haploid and diploid cells is the number of chromosome sets found in the nucleus. Diploid cells have two full sets of chromosomes and haploid cells only have one. Two haploid cells combine to make a zygote in sexual reproduction. Sexual reproduction is beneficial to offspring because it creates new and unique chromosome combinations through the joining of two haploid cells.

Haploid cells are the product of meiosis. Meiosis is a process where a diploid cell in the ovaries or testes duplicates its genes and then splits into two diploid cells. Following the first split, the cells then split again to create a total of four genetically unique haploid cells. In the reproduction process, each parent provides one haploid cell to the offspring.

Genes from the same species contain many similar characteristics but are not completely identical, or else no genetic diversity would exist within species. These variant versions of genes are known as alleles. Many alleles can exist for any gene among a species. The process of sexual reproduction breaks up alleles and generates new ones. This change in genes is beneficial to the offspring because not only does it allows them to adapt for survival in changing environmental conditions, it also accelerates the elimination of mutations from previous generations.

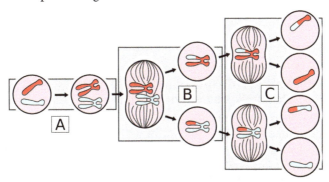

Figure 6.1 The process of meiosis.

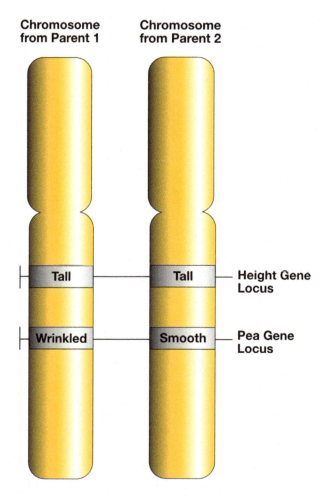

Figure 6.2 Alleles are variant versions of genes that are located in the same position, or locus, on the sister chromatids. Alleles can be the same (homozygous) or different (heterozygous). This pea plant is homozygous for height and heterozygous for skin texture.

6.1.1 Mendelian (Autosomal) Genetics

It is impossible to discuss genetics without referencing the work of Gregor Mendel, the father of genetics. His work centered on pea plants. He bred together various plants with differing characteristics and observed their offspring. The results of his studies on peas apply to all sexually reproducing organisms. Peas were chosen by Mendel because they are easy to breed and quick to reproduce in large numbers in a small space—such as the abbey garden in the monastery where he lived as a monk. His experiments unveiled the existence of dominant and recessive alleles as well as rules and ratios applicable to sexual reproduction and inheritance of genes.

Concepts in Genetics, Biochemistry, and Statistics 125

Figure 6.3 Gregor Mendel, father of genetics.

Figure 6.4 A sketch of a pea plant.

6.1.1.1 Rules of Inheritance

The collection of all genes that an individual possesses is known as their genotype. The physical expression of the genotype, the appearance of the organism, is known as its phenotype. Since each diploid cell in an organism contains two copies of each gene, one on each chromosome supplied from each parent, the alleles can either be identical or different. Two identical alleles are known as homozygous for that gene, and two different alleles are known as heterozygous. Due to genetic diversity and crossing over, the majority of the genes in humans are heterozygous.

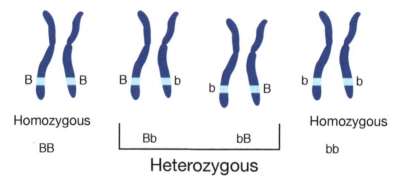

Figure 6.5 Homozygous and heterozygous chromosomes.

Mendel's studies resulted in two laws of heredity. The first is known as the law of segregation, which states that two alleles for each trait separate while gametes are being formed, and then unite again randomly at fertilization. In other words, if two yellow pea plants with one allele for yellow peas (the dominant allele, Y) and one allele for green peas (the recessive allele, y) cross-pollinate, half of the gametes from each will contain the yellow allele (Y) and half will contain the green allele (y). Therefore, four different allele combinations will exist when these gametes come together. One-quarter of the offspring will have two green pea alleles (yy), one-quarter will have two yellow pea alleles (YY), and one-half will have one yellow and one green pea allele (Yy). Since the yellow allele is dominant, three-quarters of the offspring will be yellow, and one-quarter will be green, creating a 3:1 ratio.

Mendel's second law is the law of independent assortment, which states that alleles will segregate independently during gamete formation. This was determined through the cross fertilization of two peas with two separate phenotypes: one pea contained the two dominant alleles of round shape and yellow pea color (RRYY). The other contained the two recessive alleles of wrinkled shape and green pea color (rryy). This combination produced offspring displaying a 9:3:3:1 phenotypic ratio: nine round,

Concepts in Genetics, Biochemistry, and Statistics

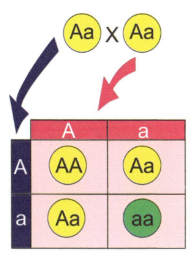

Figure 6.6 A Punnett square showing the offspring of two heterozygous pea plants.

yellow peas (RRYY, RrYY, RRYy, or RrYy), three wrinkled, yellow peas (rrYY or rrYy), three round, green peas (RRyy or Rryy), and one wrinkled, green pea (rryy).

Beyond Mendel's two laws, some more general principles of genetics are true for reproduction. Genes assort independently according to Mendel's laws, but some genes reside very close together on the same chromosome and are therefore inherited together. These "linked genes" do not assort independently. A forensic example of this are the locations on the Y chromosome. The Y chromosome is passed down the paternal line in its entirety, and so the locations examined on the Y chromosome do not assort independently. Homologous chromosomes can exhibit crossing over during meiosis, splitting up genes that are separated by some distance on the same chromosome and resulting in a recombination of the genes.

Mendel's laws apply to all autosomal nuclear DNA, but as discussed below, they do not apply to Y-chromosomal genes that are inherited paternally or to mitochondrial genes that are inherited maternally in their entirety.

6.1.1.2 Human Pedigrees

The easiest method for predicting the possible offspring from two individuals is through the use of a Punnett square. The gene or genes of interest from one parent are aligned on the top above the boxes, and the other parent's genes along the left side. Inside the boxes are the combinations of the two parent genes that intersect from that row and column, showing possible genotypes for the offspring. Punnett squares are based on Mendelian ratios of dominant and recessive alleles.

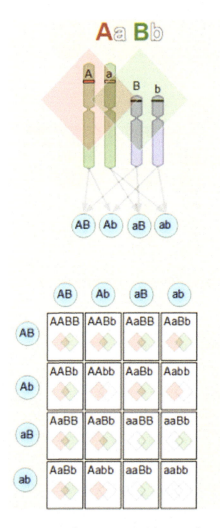

Figure 6.7 Two heterozygous chromosomes also combine to form a 9:3:3:1 phenotypic ratio.

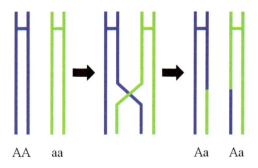

Figure 6.8 Crossing over of alleles during meiosis creates genetic diversity in the offspring.

Concepts in Genetics, Biochemistry, and Statistics

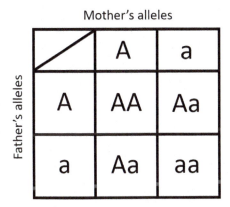

Figure 6.9 A Punnett square, showing the possible offspring from the combination of alleles from both parents.

While Punnett squares explore the possibilities of inheritance for the next generation, the best way to track a certain gene through multiple generations (past and future) is through pedigrees. Pedigrees show the phenotype for each family member following a specific trait and are useful in predicting the possibility and probability of the trait being passed on to future generations. In pedigrees, squares represent males, and circles represent females. Members that contain the gene of interest are typically shaded in to indicate its presence in that individual. Forensically, pedigrees

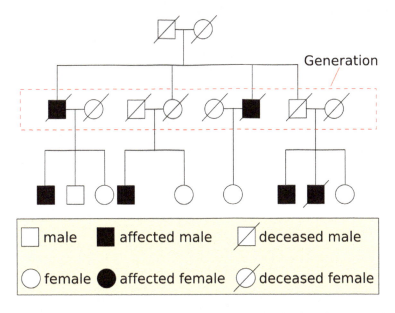

Figure 6.10 A pedigree showing the progression of color blindness in a family line.

can be helpful in tracing maternal lineage for mitochondrial DNA testing and paternal lineage for Y-STR (Y-Short Tandem Repeat) DNA testing. Pedigrees are also becoming more prominent in forensics with the rise of Forensic Genetic Genealogy.

6.1.2 Non-Mendelian

Certain DNA does not obey the Mendelian rules. With Mendel and his pea plans, each gene was either fully dominant or fully recessive, which made for straightforward results. However, human genetics are more complex than pea plant genetics. Some situations arise with the combination of alleles that go beyond dominant and recessive. Codominance is an occurrence where the two alleles for a gene are both equally expressed in the phenotype. An example of this would be a red flower and a white flower breeding and creating a flower with some red petals and some white petals.

Figure 6.11 A flower displaying codominance.

Incomplete dominance happens when the phenotype of the offspring is a combination of the two parent alleles, such as a red flower and a white flower creating a pink flower.

Two predominant cases for non-Mendelian DNA often studied in forensics include the Y-chromosomal DNA and mitochondrial DNA. They are considered non-Mendelian because the way they are passed on from parent to child is not a straightforward, equal inheritance from both parents. Both

Concepts in Genetics, Biochemistry, and Statistics 131

Figure 6.12 A flower displaying incomplete dominance.

of these types of DNA are extremely useful in many forensics settings for various reasons attributed to each.

6.1.2.1 Y-Chromosomal Inheritance

The Y chromosome is the sex chromosome specific to males, which is passed from father to son in a pattern known as patrilineage. The Y chromosome in humans holds around 50–60 genes in the form of an estimated 59 million base pairs. Y chromosomes consist of two regions: the pseudoautosomal region (PAR) and the male-specific Y region (MSY).

Pseudoautosomal regions exist on both the X and Y chromosomes. Two PARs exist on each: PAR1 and PAR2. PARs are homologous nucleotide

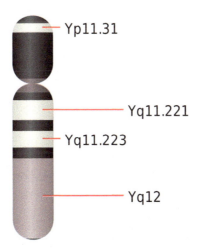

Figure 6.13 Y-chromosome with a few of its known genes labeled.

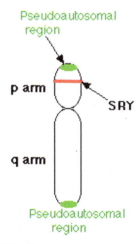

Figure 6.14 Pseudoautosomal regions are located on each end of the chromosome. The remainder of the chromosome is known as the male-specific Y region.

sequences that have a role in the segregation of the sex chromosomes in the process of meiosis. In other words, PARs help the Y chromosome find the X chromosome and possibly cross over to allow for recombination within the PAR regions of the X and Y chromosomes, to produce some male offspring with some characteristics of the mother, and female offspring with some characteristics of the father. The PAR1 region plays a more significant role since it is much larger than PAR2, so much so that if the PAR1 region is deleted on the Y chromosome it will lead to infertility in the male.

The male-specific Y region includes everything other than the PAR regions. This region is especially useful for the forensic DNA profiling testing of Y-STRs. In sexual assault cases with male suspects where the majority of the evidence comes from the DNA gathered from a female victim, the Y-STR system is extremely beneficial. Y chromosome–specific loci found in the nonrecombining region (NRY) of the Y chromosome allow for the separation of female DNA from male DNA. Y-STR loci testing can also determine patrilineage and is therefore used often for paternity testing and missing persons identification. The limit to Y-STR testing is that it cannot distinguish different individuals from the same patrilineage unless there has been some recent mutation. Though Y-STRs have a mutation rate slightly higher than what is expected in STRs, related individuals would be expected to have the same Y-STR profile for several generations of paternal descendants.

Though there are a number of Y-STR loci that have been studied, the most common amplification kits on the market type between 17 and 27 loci. Studies have shown that the analysis of a core set of Y-STRs, the European minimal haplotype standard (DYS19, DYS385-I and -II, DYS389-I and -II,

Concepts in Genetics, Biochemistry, and Statistics

DYS390, DYS391, DYS392, DYS393), can discriminate between most of the male individuals in a population. The Scientific Working Group on DNA Analysis Methods (SWGDAM) recommends utilizing a Y-STR panel that includes the European minimal haplotype plus DYS438 and DYS439. Other loci that have traditionally been included in Y-STR test kits in the United States are DYS437, DYS448, DYS456, DYS458, DYS635 (Y GATA C4), and Y GATA H4. Today, expanded STR kits include several Y-STR loci that are typed simultaneously with the STR loci for further discrimination potential, analysis quality control, and troubleshooting.

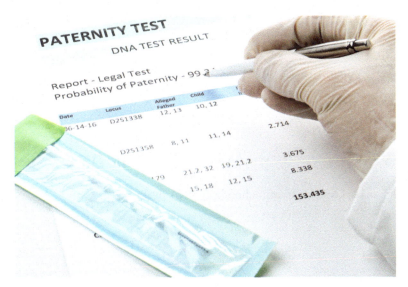

Figure 6.15 Paternity test chart.

6.1.2.2 Mitochondrial DNA

Mitochondrial DNA, or mtDNA, as mentioned in Chapter 5, is the only DNA in eukaryotic cells that exists outside of the nucleus. A mtDNA profile is knows as a mitotype. MtDNA is found in the mitochondria, which are located in the cytoplasm of the cell.

Hundreds of mitochondria exist in each cell. They produce the energy for the cell by creating high-energy molecules such as adenosine triphosphate (ATP) via aerobic metabolism. MtDNA exists in a circular loop with a control region, or displacement loop, which contains the replication origin. Human mtDNA contains 16,569 base pairs, which make up 37 genes. These genes provide codes for proteins in the respiratory complex in the mitochondria and noncoding RNA molecules that express the mitochondrial genome.

Unlike nuclear DNA, mtDNA is inherited maternally, from the mother to the child. The mitochondria contained in a sperm cell are found at the mid

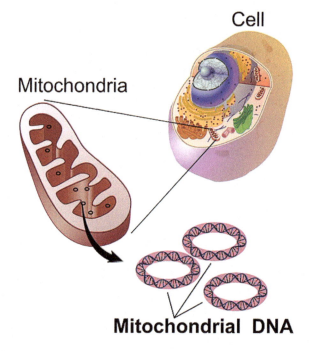

Figure 6.16 Location of mitochondrial DNA in the cell.

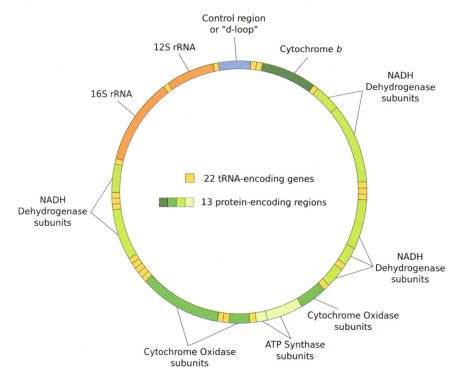

Figure 6.17 The mitochondrial genome.

Concepts in Genetics, Biochemistry, and Statistics

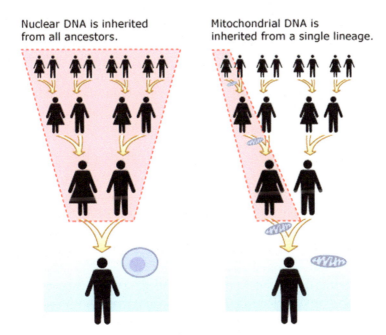

Figure 6.18 Nuclear DNA passes from every individual into the next generation, while mitochondrial DNA is only passed on by the mother.

piece, which doesn't enter the egg upon fertilization. Thus the fertilized egg only contains the maternal mitochondria, which are then passed on to all offspring. On the off chance a paternal mitochondrion enters the egg, the egg will destroy the mitochondrion after fertilization.

Many copies of mtDNA exist in each cell. This form of DNA can be useful in finding a DNA profile from samples with limited or degraded nuclear DNA. For example, a loose hair found at a crime scene or the trunk of a suspect's car that has little or no epithelial tissue attached to the root can be analyzed using mitochondrial DNA analysis with success. It is also used for victim identification in cases of missing persons or mass casualties when the only sample available for testing is human bone.

Occasionally, individuals may have a mixture of multiple mtDNA sequences, known as heteroplasmy. This is seen in hair follicles, possibly due to a merging of mtDNA from keratinocyte- and melanocyte-derived mitochondria. Melanocytes are melanin producers and have a hand in the determination of hair color. The melanosome, an organelle containing melanin, transfers itself to keratinocytes along with the melanocyte mitochondria. This cross over combines the two types of mitochondrial DNA to create a heterogeneous pool of mtDNA in hairs. On another note, individuals can display different mitotypes in different tissues of the body because mitochondrial DNA has a mutation rate significantly higher than those seen in STRs and Y-STRs.

When extracting mtDNA, different methods are used based on the sample available. Bone and teeth are cleaned and ground for easier extraction. If there is sufficient sample available, it is recommended that duplicate extractions be performed. Mitochondrial DNA testing is significantly more sensitive than traditional STR testing due to the large number of mitochondria that exist in each cell. Because this method is so much more sensitive, laboratories that employ mtDNA methods routinely have different sections of the lab reserved for this type of testing.

Prior to sequencing, the mtDNA is screened through assays such as the allele-specific oligonucleotide (ASO) assay through which suspects are often excluded or eliminated from a case. For mtDNA sequencing, the analysis of both strands of the mtDNA in a given region must be performed to ensure accuracy. Contamination should be monitored through the use of controls such as reagent blanks, negative controls, and positive controls. Sequencing of mtDNA typically happens via "PCR amplification, DNA sequencing reactions, separation using electrophoresis, and data collection and sequence analysis."[1] (Li, Forensic Biology 2nd Ed 23.3.3.). These techniques along with others will be explained in greater detail in the following chapters.

Interpreting the mtDNA profiling results includes examining the electropherogram peaks and comparing them with the suspects or family members and marking each as excluded, cannot exclude, or inconclusive. The Cambridge Reference Sequence (CRS) is the mitochondrial DNA sequence first typed in 1981. When the sequence was retyped, several errors were discovered in the original work, and the sequence became known as the revised Cambridge Reference Sequence (rCRS). When mitochondrial DNA sequencing is used for genealogical purposes, the results are often reported as differences from the rCRS. This was used as a basis for comparison with mtDNA

Figure 6.19 Electropherogram peaks, which correspond to alleles at five different loci.

Concepts in Genetics, Biochemistry, and Statistics

test results until it was replaced with the Reconstructed Sapiens Reference Sequence (RSRS).

The primary limitation of mtDNA is that it cannot distinguish individuals in the same maternal line. Mitochondrial DNA testing is also very time-consuming, and therefore if nuclear DNA is available for analysis, it is preferred. Moreover, mtDNA has a much higher mutation rate than nuclear DNA, as much as ten times higher, which can cause discrepancies when comparing mtDNA from a victim to that belonging to relatives of the victim.

6.1.3 Cytogenetics: Chromosomal Abnormalities

Cytogenetics, or the study of chromosomes, has evolved rapidly over the past few decades in the area of recognizing abnormalities on chromosomes. Despite the self-correcting and proofreading mechanisms set in place by the cell, chromosomal abnormalities may arise over the course of one's life through complications in mitosis, as well as possible abnormalities in the next generation through complications in meiosis.

6.1.3.1 Nondisjunction: Mitotic and Meiotic

The processes of mitosis and meiosis are not flawless. Sometimes, chromosomes do not separate properly in an event known as nondisjunction. In mitosis, this causes one daughter cell to have both copies of a certain chromosome, and the other to have no copies, both of which are genetically unstable. In meiosis, nondisjunction causes some of the haploid cells produced to contain either too many or too few chromosomes. This is known as aneuploidy.

If these abnormal gametes become fertilized, the embryos produced are very likely to not make it to full development and will die in the womb. However, some abnormal gametes will survive. Trisomy is a type of polysomy where there are three copies of a particular chromosome, instead of

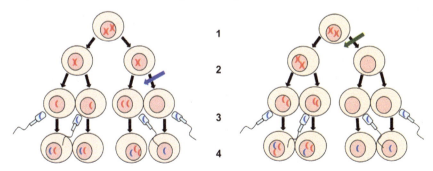

Figure 6.20 Two examples of nondisjunction. The first example shows nondisjunction in meiosis II, causing aneuploidy in two of the four gametes. The second example shows nondisjunction in meiosis I, causing aneuploidy in all four gametes.

the normal two. Trisomy is a type of aneuploidy that can result from nondisjunction. The most common type of autosomal trisomy is trisomy 21 (Down syndrome), which is caused by an extra copy of chromosome 21 that occurs during nondisjunction in meiosis I. The resulting gamete contains three copies of the 21st chromosome instead of the regular two, therefore producing an extra dose of the proteins that are encoded on that chromosome, interfering with the development of the embryo and some functions in the adult. Other common types of autosomal trisomy are trisomy 18 (Edwards syndrome), trisomy 13 (Patau syndrome), trisomy 9, and trisomy 8 (Warkany syndrome 2). These can sometimes be visible in forensic DNA typing when there are triallelic patterns at D21, D18, D13, D9, and D8. Trisomy can also be seen in sex chromosomes.

Nondisjunction is a more common phenomenon than one might think—in females, nondisjunction occurs in approximately 10% of meiosis cycles. Nondisjunction is less common in males since sperm development contains a quality control step where an improper meiosis cycle activates apoptosis, or cell death. Nondisjunction is believed to be the root of the miscarriage rate in early pregnancy.

6.1.3.2 *Chromosomal Abnormalities*

Mutations in chromosomes affect the proteins produced in many ways. Mutations that reduce or eliminate gene activity are known as loss-of-function mutations. These mutations are typically recessive, decreasing the amount of gene product normally produced by 50%, which has little impact on the individual. In contrast, gain-of-function mutations increase the activity of the gene product, and unfortunately these mutations are frequently dominant. An example of a gain-of-function mutation is the commonly mutated *Ras* gene, which plays a role in controlling cell division. Mutation on this gene produces an overabundance of signals for cells to divide, which promotes cancer development.

6.1.4 Genetic Disease

Certain diseases are genetically predisposed. The inevitability of the disease is unfortunate, but for scientists, predisposed diseases are at times easier to treat since finding the gene or genes responsible for the disease can help predict the diagnosis at an early stage as well as give insight into treatment and prevention. Finding the gene associated with a disease is made possible through access to single nucleotide polymorphisms (SNPs). SNPs are occurrences of single nucleotides that are common to the entire human race. More than 17 million SNPs have been identified and cataloged. SNPs are markers of interest to the forensic community because of their abundance in the human genome, their relatively low mutation rate, and the possibility of automating the analysis with high-throughput technologies. Moreover, SNPs are smaller DNA fragments than STRs, which may prove useful when attempting to obtain a DNA profile from degraded DNA samples.

Many new technologies for typing SNPs have been developed in the past few years, such as massively parallel sequencing (MPS). The term "Massively Parallel Sequencing" is used to describe the method of high-throughput DNA sequencing to determine the entire genomic sequence of a person or organism. This method processes millions of reads, or DNA sequences, in parallel instead of processing single amplicons that generate a consensus sequence. Also, next-generation sequencing (NGS) refers to non-Sanger-based high-throughput DNA sequencing technologies. Millions or billions of DNA strands can be sequenced in parallel, yielding substantially more throughput and minimizing the need for the fragment-cloning methods that are often used in traditional Sanger sequencing of genomes.

By knowing what is common among humans, scientists can identify the differences that create certain genetic disorders from issues such as obesity, diabetes, asthma, and arthritis. This comparison is done through the creation of a genetic linkage map, which displays two alleles side by side to discover the discrepancies and similarities. Many mutations that harm the activity of a major gene will typically have disastrous effects on the individual, thus causing them to not have offspring, which causes the mutation to eliminate itself via natural selection. It is the genetic problems that only alter gene function slightly and not to a lethal level that are more common and tend to be passed on. In order to broaden the search for genetic variables that can affect the risk for common diseases, genome wide association studies can be performed.

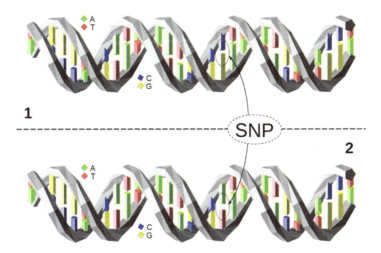

Figure 6.21 DNA strands from two different individuals showing a single nucleotide difference.

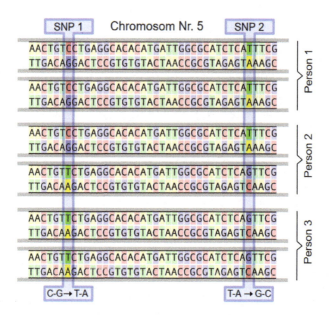

Figure 6.22 DNA from three individuals with three different profiles resulting from differences in only two SNP loci.

Note

1. Li, R. C. (2015). *Forensic biology* (2nd ed.). Boca Raton: CRC Press.

Bibliography

Alberts, B., Bray, D., Hopkin, K., Johnson, A., Lewis, J., Raff, M., & Roberts, K. (2014). *Essential cell biology* (4th ed.). New York: Garland Science (Ch 19, 20, 25).
Li, R. C. (2015). *Forensic biology* (2nd ed.). Boca Raton: CRC Press (Ch 3, 4, 21, 23).
Saferstein, R. (2018). *Criminalistics: an introduction to forensic science* (12th ed.). Boston: Pearson (Ch 10).

Concepts in Genetics, Biochemistry, and Statistics cont.

7.1 Population Genetics

The gene pool within a species contains wide variation. Massive quantities of alleles vary vastly throughout a population. The study of the patterns of genetic variation and their causes within populations is known as population genetics. While alleles at some loci are simple and code for one or two variants, like having the ability to curl your tongue or not, some alleles are more diverse, including codes for dominant, recessive, and codominant variations. The probability of certain genetic combinations being passed on in controlled populations can be calculated through basic statistics that have been described by two men: G. H. Hardy and Wilhelm Weinberg.

Figure 7.1 G. H. Hardy.

Figure 7.2 Wilhelm Weinberg.

7.1.1 Hardy–Weinberg

Hardy–Weinberg equilibrium states that allele and genotype frequencies in a population will remain constant from generation to generation in the absence of other evolutionary influences, such as gene flow, genetic drift, and mate choice. The principle discusses the allele to genotype frequency in a population while at equilibrium. Allele frequency is found by counting one type of allele at a given locus and dividing that by the total number of alleles at that locus. Genotype frequency is found by counting the number of individuals with one genotype and dividing that by the total number of individuals in the population. The sum of all genotype frequencies should be 1. The study of alleles at a given locus can show homozygosity or heterozygosity. Heterozygosity is the proportion of alleles that are heterozygous at a given locus, which speaks to the amount of genetic variation in the sample population. High heterozygosity indicates high levels of variation at a given locus.

The Hardy–Weinberg principle hinges on a few assumptions: that allele frequencies will not change between generations, and that genotype frequencies can be predicted based on allelic frequencies. To ensure that allele frequencies do not change between generations, in a sample population there must be random mating, no selection, no mutations, a large population, and no migration. The concepts of random mating and no selection imply that neither natural nor unnatural selection is taking place that would change allele frequencies. No mutations mean that the dominant and recessive alleles remain the same, and don't transform into a new and unique trait. A large population is used in order to eliminate anomalies in the population so the results are as accurate as possible. No migration ensures that no new alleles are introduced and no alleles leave. Essentially, these conditions ensure that stable allele frequency is maintained in the population.

Concepts in Genetics, Biochemistry, and Statistics cont. 143

The second assumption made in the Hardy–Weinberg principle is that genetic frequencies are predicted based on allelic frequencies. The calculation to predict genetic frequencies takes place with the Hardy–Weinberg equation. This equation relates to the percentages of dominant and recessive alleles. This implies that there are only two alleles for a certain trait, one dominant and one recessive. For the purposes of better explaining the equation, we will call the dominant allele B to represent brown hair, and the recessive allele b to represent blond hair. The percentage of the dominant allele (B) in the population is denoted with a p, and the percentage of the recessive allele (b) with a q.

$$p+q=1$$

Naturally, all of the population with the dominant trait added to all of the population with the recessive trait is equal to 100% of the population. Taking this equation further, both sides are squared to produce the equation

$$(p+q)^2=(1)^2$$

$$p^2+2pq+q^2=1$$

It is helpful to think of p as the probability of randomly choosing a dominant allele out of the total population, and likewise to think of q as the probability of randomly choosing a recessive allele out of the total population. Therefore, p^2 is the probability of choosing someone with two dominant alleles, who is homozygous dominant (BB). By the same logic, q^2 is the probability of someone having two recessive alleles, who is homozygous recessive (bb). The figure 2pq is a little more complicated to explain; imagine that you select a person who has a brown (B) gene from their father and a blond (b) gene from their mother. They represent pq, a combination of both alleles. Moreover, what if that person received the brown (B) gene from their mother and the blond (b) gene from their father? That also qualifies as pq, which is why the factor has a coefficient of 2 in the equation. In other words, there are two ways of becoming a heterozygote (Bb), (p × q), or (q × p). It makes sense that since all of the population is either homozygous dominant (BB), heterozygous (Bb or bB), or homozygous recessive (bb), the three terms add to 100%.

It is important to remember that Hardy–Weinberg is a discussion of genotypes, not phenotypes. However, in some cases, if you are presented with information on the phenotypes within a population, this can be utilized to find the genotype quantity as well. For example, if it was found that 9% of the population had blond hair, it would likewise mean that 9% of the population is homozygous recessive (bb), since that is the only genotype that will produce a phenotype of blond hair. If q^2 = 9%, or 0.09, then q = $\sqrt{0.09}$, which

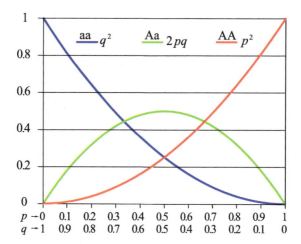

Figure 7.3 A graph relating the quantities of each allele combination in the population.

equals 0.3 or 30%. Since p + q = 1, therefore p = 1 − q, or 1 − 0.3. Therefore, p = 0.7, or 70%.

$q^2 = 0.09$	$q = 0.3$
$p^2 = 0.49$	**p = 0.7**

To find the percentage of homozygous dominant genotypes, you square p: $0.7^2 = p^2 = 0.49$, or 49%. Plugging p^2 and q^2 into the longer Hardy–Weinberg equation tells us that 2pq = 0.42, or 42%. In other words, 49% of the population is homozygous dominant (BB), 42% is heterozygous (Bb), and 9% is homozygous recessive (bb).

If a population's allele frequencies align with the Hardy–Weinberg principles, it is said to be in Hardy–Weinberg equilibrium. To test for equilibrium, *observed* genotype frequencies are collected at a given locus by dividing the individuals with one genotype by the total number of individuals in the population. *Expected* genotype frequencies are calculated using the equation $p^2 + 2pq + q^2 = 1$ as described above. Lastly, observed and expected genotype frequencies are entered into a chi-square test to determine the significance of the difference between the two groups. Chi-square (x^2) is calculated with the following equation:

$$x^2 = \sum_{i=1}^{n} \frac{(O_i - E_i)^2}{E_i}$$

O_i represents the observed genotype frequency
E_i represents the expected genotype frequency
n represents the total number of genotypes.

Degrees of freedom can be calculated from the number of genotypes minus the number of alleles. From the degrees of freedom and the chi-square

Concepts in Genetics, Biochemistry, and Statistics cont.   145

value a p value (different to the letter p above, which represents dominant allele frequency) is found from a table. A p value larger than 0.05 indicates that there is not a significant difference between observed and expected genotype frequencies, and the null hypothesis that the observed and expected are not significantly different is accepted. In the forensics world, Hardy–Weinberg is useful for calculating the population match probability, or the probability of two randomly chosen individuals having matching genotypes. When doing DNA analysis of a suspect or victim's DNA alongside the DNA from a piece of evidence, it is never desirable to find a match between two random individuals and accidentally connect the wrong suspect or victim to evidence. Population match probability (P_m) can be calculated by the following formula, where p and q again relate to the percentages of the dominant and recessive alleles in the population:

$$P_m = \sum_i (p^2)^2 + \sum_i (2pq)^2$$

7.1.2 Mechanisms of Evolution

The basis of the Hardy–Weinberg theory rests on the maintenance of stable allele frequency. Evolution via mechanisms such as mutation or selection makes the Hardy–Weinberg principle inapplicable, but they are realities of the world and must therefore be taken into account. Evolution, which simply denotes change over time, is caused by a violation of the Hardy–Weinberg assumptions, namely mutation, non-random mating, genetic drift, gene flow via migration, and natural selection.

7.1.2.1 Mutation

A mutation is the changing of the structure of a gene, resulting in a variant form that may be transmitted to subsequent generations. Mutations can be caused by the alteration of single-base units in DNA, or the deletion, insertion, or rearrangement of larger sections of genes or chromosomes. From one generation to the next, mutations don't typically have a large impact. Mutations can have a large effect when combined with natural selection, in that an individual may no longer be as likely to be selected for mating if they have a mutation, which can drive evolution away from that particular mutation.

7.1.2.2 Selection

Selection implies the choosing of a mate based on their phenotype or genotype: either a similar partner or a different one. In Hardy–Weinberg, random mating implies that phenotypes and genotypes are not taken into account when mating, but in reality that is not the case. Certain physical disabilities,

146 Guide to the ABC Biology Exam

mental disorders, or genetic predispositions may prevent that organism/individual from being selected as a mate. Non-random mating can therefore alter the frequencies of genotypes in the next generation and those to follow.

7.1.3 Statistics and Probability

The Mendelian inheritance of genes can often be measured using probabilities. A probability is the ratio of the number of actual occurrences of an event to the number of possible occurrences. In short, it is an estimation of how common or rare an event is expected to be using a sampling of the population of interest. When two independent events occur, such as the inheritance of two different alleles at a locus, the probability of the two independent events occurring simultaneously is the product of each of their individual probabilities. This is known as the product rule of probability. Statistical analysis for genotypes in a non-Hardy–Weinberg world can be performed through the use of likelihood ratios and calculation of profile probability.

7.1.3.1 Likelihood Ratios

Likelihood ratios evaluate an evidence DNA profile and compare the likelihood of two competing hypotheses. Hypothesis 1 (H_1), also referred to as the prosecutors' hypothesis (H_p), states that the DNA profiles obtained from the evidence and the suspect came from the same source. Hypothesis 2 (H_2), also called the defense hypothesis (H_d), states that the DNA profiles obtained from the evidence and the suspect did not come from the same source and some random unrelated person is actually the source. The likelihood ratio (LR) is actually a ratio of two probabilities. It's the probability of observing the DNA typing results if H_1 is true divided by the probability of observing the DNA typing results if H_2 is true. An LR can be calculated by dividing the probability of H_1 by the probability of H_2:

$$LR = \frac{Pr_{H1}}{Pr_{H2}}$$

The larger the numerator, the larger the likelihood ratio, which supports the proposition that the evidence could have resulted from the suspect. This is usually expressed in whole numbers greater than 1; the larger the number the stronger the support for inclusion. When a match between suspect and evidence occurs, Pr_{H1} equals 1 (or 100%), and Pr_{H2} equals the profile probability. A likelihood ratio of 1 is considered neutral, no support for either hypothesis over the other. The larger the denominator, the smaller the likelihood ratio, which supports the proposition that the evidence could have resulted from some unknown, unrelated individual. Likelihood ratios less than 1 support exclusion and the smaller the decimal number, the stronger the support for exclusion.

Concepts in Genetics, Biochemistry, and Statistics cont.

7.1.3.2 *Pd and Pi*

As discussed above, allele frequencies can be calculated using the counting method, which involves counting the number of alleles of one type at a given locus and dividing it by the total number of alleles at that locus in a sampled population. Genotype frequencies are calculated by dividing the number of individuals with one genotype by the total number of individuals in a sampled population. The sum of all the genotype frequencies at any given locus should equal 1. Each genotype will have its own individual frequency and the sum of the genotype frequencies of all the loci tested is the profile probability. A low profile probability indicates that an individual randomly chosen is less likely to match a DNA profile taken from a piece of evidence. Profile probability is calculated by first finding the locus genotype frequencies and then multiplying them using the product rule. Locus genotype frequency for homozygotes is $P_i = p^2$. Locus genotype for heterozygotes is $P_i = 2pq$. These formulas assume that a person randomly selected is not related to the perpetrator of the crime.

A necessary correction factor must be inserted with profile probabilities to take into account the effect of population structure. This correction factor is denoted as θ, also referred to as the inbreeding coefficient. The value of θ for the majority of the U.S. population is 0.01, and 0.03 for Native Americans and other isolated populations. With this correction, profile probability decreases by about three times. This is the corrected formula for locus genotype frequency in homozygotes:

$$P_i = \frac{[2\theta + (1-\theta)p][3\theta + (1-\theta)p]}{(1+\theta)(1+2\theta)}$$

The corrected formula for locus genotype frequency in heterozygotes is as follows:

$$P_i = \frac{2[\theta + (1-\theta)p][\theta + (1-\theta)q]}{(1+\theta)(1+2\theta)}$$

7.1.3.2.1 Population Substructure—Calculations—NRC II

In the early 1990's, forensic DNA profiling techniques were being fine-tuned for use in casework. The National Research Council (NRC) of the National Academy of Sciences published a report in 1994 with recommendations and possible ethical red flags in the area of statistics as applied to DNA testing. The report wasn't well received, and the NRC published a second revised report in 1996 known as NRC II. The goal of this report, which focused on statistical calculations applied to forensic DNA profiles, was to understand the benefits and limitations of forensic DNA profiling in assisting a judge or jury in the process of drawing conclusions in a case.

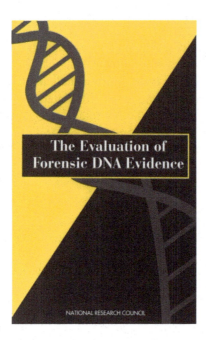

Figure 7.4 The NRC II.

7.1.4 Population Databases

As was described and recommended in the NRCII, population databases were constructed to demonstrate how common or rare a given DNA profile is expected to be in the population. There have been a number of databases that have been constructed and published. The most widely used forensic DNA frequency databases in the United States are the FBI database, first published in 1999, and the NIST database, published in 2012. Some laboratories have created their own databases, which they have published. In these databases, a statistically representative sample of population (approximately 100–200 samples per subgroup) is collected and analyzed. The allele frequencies in the population database are used by the laboratory to calculate genotype frequencies at each locus in a given DNA profile using the standard equations p^2, $2pq$, and q^2. Profile probability can then be calculated using the product rule.

Calculation of profile probability is often reported as: The probability of a randomly selected, unrelated individual also matching the profile observed is 1 in every ___ individuals.

Concepts in Genetics, Biochemistry, and Statistics cont.

Allele	CSF1PO	FGA	TH01	TPOX	VWA	D3S1358	D5S818	D7S820	D8S1179	D13S317	D16S539	D18S51	D21S11	D2S1338	D19S433
5	--	--	0.002	0.002	--	--	--	--	--	--	--	--	--	--	--
6	--	--	0.232	0.002	--	--	--	--	--	--	--	--	--	--	--
7	--	--	0.190	--	--	--	0.002	0.018	--	--	--	--	--	--	--
8	0.005	--	0.084	0.535	--	--	0.003	0.151	0.012	0.113	0.018	--	--	--	--
8.1	--	--	--	--	--	--	--	0.002	--	--	--	--	--	--	--
9	0.012	--	0.114	0.119	--	--	0.050	0.177	0.003	0.075	0.113	--	--	--	--
9.3	--	--	0.368	--	--	--	--	--	--	--	--	--	--	--	--
10	0.217	--	0.008	0.056	--	--	0.051	0.243	0.101	0.051	0.056	0.008	--	--	0.002
10.3	--	--	--	--	--	--	--	--	--	--	--	--	--	--	--
11	0.301	--	0.002	0.243	--	0.002	0.361	0.207	0.083	0.339	0.321	0.017	--	--	0.005
12	0.361	--	--	0.041	--	--	0.384	0.166	0.185	0.248	0.326	0.127	--	--	0.081
12.2	--	--	--	--	--	--	--	--	--	--	--	--	--	--	0.002
13	0.096	--	--	0.002	0.002	--	0.141	0.035	0.305	0.124	0.146	0.132	--	--	0.253
13.2	--	--	--	--	--	--	--	--	--	--	--	--	--	--	0.007
14	0.008	--	--	--	0.094	0.103	0.007	0.002	0.166	0.048	0.020	0.137	--	--	0.369
14.2	--	--	--	--	--	--	--	--	--	--	--	0.002	--	--	0.018
15	--	--	--	--	0.111	0.262	0.002	--	0.114	0.002	--	0.159	--	0.002	0.152
15.2	--	--	--	--	--	--	--	--	--	--	--	--	--	--	0.035
16	--	--	--	--	0.200	0.253	--	--	0.031	--	--	0.139	--	0.033	0.050
16.2	--	--	--	--	--	--	--	--	--	--	--	--	--	--	0.015
17	--	--	--	--	0.281	0.215	--	--	--	--	--	0.126	--	0.182	0.008
17.2	--	--	--	--	--	--	--	--	--	--	--	--	--	--	0.002
18	--	0.026	--	--	0.200	0.152	--	--	--	--	--	0.076	--	0.079	--
18.2	--	--	--	--	--	--	--	--	--	--	--	--	--	--	0.002
19	--	0.053	--	--	0.104	0.012	--	--	--	--	--	0.038	--	0.114	--
19.2	--	--	--	--	--	--	--	--	--	--	--	--	--	--	--
20	--	0.127	--	--	0.005	0.002	--	--	--	--	--	0.022	--	0.146	--
21	--	0.185	--	--	0.002	--	--	--	--	--	--	0.008	--	0.041	--
21.2	--	0.005	--	--	--	--	--	--	--	--	--	--	--	--	--
22	--	0.219	--	--	--	--	--	--	--	--	--	0.008	--	0.038	--
22.2	--	0.012	--	--	--	--	--	--	--	--	--	--	--	--	--
22.3	--	--	--	--	--	--	--	--	--	--	--	--	--	--	--
23	--	0.134	--	--	--	--	--	--	--	--	--	--	--	0.118	--
23.2	--	0.003	--	--	--	--	--	--	--	--	--	--	--	--	--
24	--	0.136	--	--	--	--	--	--	--	--	--	--	--	0.123	--
24.2	--	0.002	--	--	--	--	--	--	--	--	--	--	--	--	--
25	--	0.071	--	--	--	--	--	--	--	--	--	--	--	0.093	--
25.2	--	--	--	--	--	--	--	--	--	--	--	--	0.002	--	--
26	--	0.023	--	--	--	--	--	--	--	--	--	--	--	0.030	--
27	--	0.003	--	--	--	--	--	--	--	--	--	--	0.026	0.002	--
28	--	--	--	--	--	--	--	--	--	--	--	--	0.159	--	--
29	--	--	--	--	--	--	--	--	--	--	--	--	0.195	--	--
29.2	--	--	--	--	--	--	--	--	--	--	--	--	0.003	--	--
30	--	--	--	--	--	--	--	--	--	--	--	--	0.278	--	--
30.2	--	--	--	--	--	--	--	--	--	--	--	--	0.028	--	--
31	--	--	--	--	--	--	--	--	--	--	--	--	0.083	--	--
31.2	--	--	--	--	--	--	--	--	--	--	--	--	0.099	--	--
32	--	--	--	--	--	--	--	--	--	--	--	--	0.007	--	--
32.2	--	--	--	--	--	--	--	--	--	--	--	--	0.084	--	--
33	--	--	--	--	--	--	--	--	--	--	--	--	0.002	--	--
33.1	--	--	--	--	--	--	--	--	--	--	--	--	--	--	--
33.2	--	--	--	--	--	--	--	--	--	--	--	--	0.026	--	--
34	--	--	--	--	--	--	--	--	--	--	--	--	--	--	--
34.2	--	--	--	--	--	--	--	--	--	--	--	--	0.005	--	--
35	--	--	--	--	--	--	--	--	--	--	--	--	0.002	--	--
36	--	--	--	--	--	--	--	--	--	--	--	--	--	--	--
37	--	--	--	--	--	--	--	--	--	--	--	--	--	--	--
38	--	--	--	--	--	--	--	--	--	--	--	--	--	--	--
39	--	--	--	--	--	--	--	--	--	--	--	--	--	--	--
H(ob)	0.725	0.887	0.719	0.656	0.841	0.765	0.709	0.818	0.778	0.745	0.735	0.881	0.841	0.871	0.755
H(ex)	0.724	0.857	0.756	0.637	0.810	0.789	0.698	0.816	0.816	0.786	0.754	0.880	0.835	0.885	0.767
P	0.968	0.037	0.058	0.522	0.202	0.628	0.891	0.423	0.278	0.099	0.171	0.846	0.205	0.798	0.952

H(ob): observed heterozygosity; H(ex): expected heterozygosity; P: Hardy-Weinberg equilibrium, exact test based on 2000 shufflings.

Figure 7.5 STR data table. (Used with permission, courtesy of John Butler.) Source: Butler, J., et al., *J Forensic Sci*, 48, July 2003.

7.2 Non-Human Molecular Applications

When forensic personnel arrive at a crime scene, they are most likely expecting to encounter evidence and DNA left from humans. While this is often the case, encounters with animal, plant, and microbial DNA evidence are not impossible, and they may be the puzzle piece that solves a case. When evidence from animals, plants, or chemicals is found at a scene, on a suspect, or in a suspect's home or car, such evidence should be carefully retained and analyzed.

7.2.1 Animal Forensic DNA Applications

When you consider how many American households have domesticated pets, it makes sense that animal samples could also become important evidence at a crime scene. DNA testing kits exist for routine STR testing for cats, dogs, and other domesticated animals. Species determination may be important for crime scenes where animals are also victims of the crime. Animal blood can be distinguished from human blood via the precipitin test, described in Chapter 4. To summarize, human and animal antiserum contains different antibodies that will react with only human or animal antigens. Commercially available antiserums to test for animal blood exist for commonly encountered animals, such as cats, dogs, and deer.

DNA analysis can be done on the animal sample by investigating common loci such as the mitochondrial cytochrome b gene, the cytochrome c oxidase I gene, and the D-loop. PCR is useful not only for determining if the DNA is human— but if the DNA is animal, 221 animal species can be recognized using different sets of primers. Specifically, the primers amplify a segment on mitochondrial cytochrome b, the results of which can identify the family, genus, and species of the sample. DNA testing is used on non-human DNA in wildlife crime investigation, breeder parentage verification, DNA genotyping, and inherited disease screening.

7.2.2 Plant Forensic DNA Applications

Forensic botany, while not a standard component in most crime scenes, can provide vital information in rare cases that may connect a perpetrator to a victim or crime scene. Sometimes plant material can be involved in a crime scene—for example, in the case of a body being transported from one

Figure 7.6 An FBI forensics team searches an outdoor crime scene to collect samples of the dirt and leaves surrounding the location of evidence found.

Concepts in Genetics, Biochemistry, and Statistics cont. 151

Figure 7.7 The Palo Verde Seed Pod was used in the homicide case of Denise Johnson to link the suspect Mark Bogan to the scene where Johnson's body was left.

Figure 7.8 Marijuana, the plant matter submitted most frequently to forensic drug labs for analysis.

location to another and their clothes retaining leaves or fibers of a plant from the area of the crime, or when plant material in a suspect's car matches a tree near where a body was found.

Similar to the primers specific to animal DNA, DNA markers exist for a wide range of plant species and can be identified using PCR amplification. Plant DNA testing can also assist in drug trafficking investigations involving propagation and grow operations for drugs such as heroin, cocaine, and marijuana.

7.2.3 Microbial DNA Applications

Microbial forensics is a relatively new discipline that investigates bioterrorism crimes or the release of microorganisms or toxins. Crimes such as these can aim to harm or kill people, animals, or crops. Microbial DNA is also present in and on the human body in abundance—approximately ten times more microbial cells than human cells. Metagenomics is a new method of DNA sequencing that allows for the study of microbial DNA harvested from its natural environment. Identification of microbes is helpful in certain cases in forensics due to the fact that microbial diversity is large, meaning that different variations of microbes exist in different areas of the body. Body fluid identification using microbial studies can aid the forensic investigation. Like animal and plant DNA, microbial DNA is identified using DNA markers specific for primer binding sites and PCR amplification. The potential forensic utility of microbial DNA has been demonstrated for samples collected from highly diverse locations, including the environment, hair, skin, and the vagina. Examples of forensically relevant information that can be obtained from microbial analysis include indications of personal contact, geographic location information, and estimations on time of death.

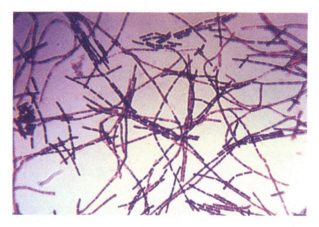

Figure 7.9 Bacillus Anthracis.

Bibliography

Budowle, B., Moretti, T. R., Baumstark, A. L., Defenbaugh, D. A., & Keys, K. M. (1999). Population data on the thirteen CODIS core short tandem repeat loci in African Americans, U.S. Caucasians, Hispanics, Bahamians, Jamaicans, and Trinidadians. *Journal of Forensic Science*, 44, 1277–1286.

Butler, J. M., Hill, C. R., & Coble, M. D. (2012). Variability of New STR Loci and Kits in US Population Groups. Retrieved from http://www.promega.com/resources/articles/profiles-in-dna/2012/variabilityof-new-str-loci-and-kits-in-us-population-groups/

Concepts in Genetics, Biochemistry, and Statistics cont. 153

Clarke, T. H., Gomez, A., Singh, H., Nelson, K. E., & Brinkac L. M. (2017). Integrating the Microbiome as a Resource in the Forensics Toolkit. Retrieved August 31, 2018 from https://www.sciencedirect.com/science/article/pii/S1872497317301400

Coble, M. D., Hill, C. R., & Butler J. M. (2013). Haplotype Data for 23 Y-chromosome Markers in four U.S. Population Groups. *Forensic Science International: Genetics, 7*, e66–e68.

Craft, K. J., Owens, J. D., & Ashley, M. V. (January 5, 2007). Application of Plant DNA Markers in Forensic Botany: genetic Comparison of Quercus Evidence Leaves to Crime Scene Trees using Microsatellites. Retrieved June 2, 2017 from https://www.ncbi.nlm.nih.gov/pubmed/16632287

Hill, C. R., Duewer, D. L., Kline, M. C., Coble, M. D., & Butler, J. M. (2013). U.S. Population Data for 29 Autosomal STR Loci. *Forensic Science International: Genetics, 7*, e82–e83.

Khan Academy. (2017). Applying the Hardy-Weinberg Equation. Retrieved June 02, 2017 from https://www.khanacademy.org/science/biology/her/heredity-and -genetics/v/applying-hardy-weinberg

Khan Academy. (2017). Discussions of Conditions for Hardy Weinberg. Retrieved June 2, 2017 from https://www.khanacademy.org/science/biology/her/heredit y-and-genetics/v/discussions-of-conditions-for-hardy-weinberg

Khan Academy. (2017). Hardy-Weinberg Equation. Retrieved June 2, 2017 from https://www.khanacademy.org/science/biology/her/heredity-and-genetics/v/ hardy-weinberg

Khan Academy. (2017). Mechanisms of Evolution. Retrieved June 2, 2017 from https ://www.khanacademy.org/science/biology/her/heredity-and-genetics/a/har dy-weinberg-mechanisms-of-evolution

Li, R. C. (2015). *Forensic biology* (2nd ed.). Boca Raton: CRC Press (Ch 11, 13, 25).

Saferstein, R. (2018). *Criminalistics: an introduction to forensic science* (12th ed.). Boston: Pearson (Ch 3, 14).

Verma, S. K., & Singh, L. (December 14, 2002). Novel Universal Primers Establish Identity of an Enormous Number of Animal Species for Forensic Application. Retrieved June 2, 2017 from http://onlinelibrary.wiley.com/doi/10.1046/j.1471-8286.2003.00340.x/full

Zaya, D. N., & Ashley, M. V. (2012). Plant Genetics for Forensic Applications. Retrieved June 2, 2017 from https://www.ncbi.nlm.nih.gov/pubmed/22419487

History and Standards of DNA Evidence

8

8.1 Types of Evidence

Physical evidence is the backbone of forensic science and can exist in many different forms. The common types of physical evidence include: blood, semen, saliva, documents, drugs, explosives, fibers, fingerprints, firearms and ammunition, glass, hair, impressions, organs and physiological fluids, paint, petroleum products, plastic bags, rubber and other polymers, powder residues, serial numbers, soil and minerals, tool marks, vehicle lights, wood, and other vegetative matter. Depending on the nature of the case, certain types of evidence are more likely to be encountered than others.

Due to the increased popularity of television shows such as CSI, the general public has developed unrealistic expectations regarding the recovery and processing of forensic evidence in criminal cases. The "CSI effect" is the unreasonable expectation by jurors that there will be forensic evidence in every criminal case. This specifically affects forensic DNA analysts and the demand for DNA testing. According to a study performed by the National Institute of Justice in 2011, 46% of people surveyed expect to see forensic evidence in every criminal case, 22% expect to see DNA evidence in every criminal case, and 73% expect to see DNA evidence in every case of sexual assault.

In truth, less than half of all criminal cases have any type of forensic evidence. The field has never been as technologically sophisticated as it is today, and the methods that are used have advanced exponentially since the inception of forensic science. Even with these advancements, useful scientific evidence will not be recovered in every case. When it is present, forensic scientists have developed cutting-edge methods to discover it. Hundreds of thousands of scientists worldwide work tirelessly to develop new techniques and technologies to improve detection and recognition of important evidence. This chapter will explain these methods for discovery and analysis of evidence as well as specifying which kind of cases each type of evidence is typically encountered in. Moreover, we will also discuss the evolution of forensic biology, as well as accepted standards and practices in the field.

Figure 8.1 *CSI: Crime Scene Investigation* is a television show that increased public interest in forensic science.

8.1.1 Criminal

All cases under the law are divided into two types: civil and criminal. Civil laws deal with situations that arise from non-criminal disputes which usually involve money. Examples of civil conflicts are divorce, child custody, eviction, and bankruptcy. Criminal laws deal with instances when someone has committed a crime. Forensic biology evidence is most commonly used in relation to criminal cases, but can certainly be used in immigration and paternity matters before the court. The individuals who are charged with collecting physical evidence that can aid in the investigation of a crime are crime scene investigators, or CSIs. They have the vital responsibility of recognizing important evidence as well as ensuring proper documentation and packaging.

Blood, semen, and saliva, from either animal or human sources, must all be sampled. These may be found in liquid form, as with fresh bloodstains, or dried onto carpet, clothing, or other objects. Saliva may be present on things like unwashed cups, cigarette butts, or undergarments in sexual assault cases.

Figure 8.2 Blood droplets found at a crime scene.

History and Standards of DNA Evidence 157

Figure 8.3 Cigarette butts are a common source of saliva evidence that may be found at a crime scene.

Figure 8.4 Used condoms can be a rich source of semen evidence that may be found at a crime scene.

Information gleaned from evidence in a criminal case should help to prove or disprove some fact of consequence. The presence of a suspect's semen in the vaginal cavity of an underage girl proves that illicit sexual contact took place. The presence of a person's saliva on a cigarette butt found at the crime scene proves that person was likely present at that location at some point in time.

8.1.2 Non-Criminal

Questions of paternity and also relatedness for the purpose of immigration are non-criminal matters, but may require the use of forensic biology evidence. If a man fathers a child, he may be obligated to provide financial support for the child over the course of the child's life. In order for the mother of the child to be awarded such support, paternity must be established by the court through DNA testing. The identities or relationships of individuals wishing to immigrate to the United States must be established to ensure proper procedures are followed. For example, parents who have been granted legal status may wish to seek legal immigration status for their minor children. In some cases, special processes exist for first-order family members of those individuals who have already been granted a legal status in the United States. In order for those exceptions to apply, the familial relationship must be established by the courts through DNA testing. This is considered a civil, non-criminal matter. Forensic biology evidence may be significant in divorce claims of unfaithfulness, and in civil suits for medical malpractice or negligence. DNA testing can be used in agriculture to confirm fish fillets imported are that of the species the vendor claims they are, and that steaks are beef. DNA testing can even be used to test dog feces in residential communities to determine who is not cleaning up after Fido does his business.

8.1.3 Missing-Person/Mass Disaster

In cases where human remains no longer exhibit recognizable characteristics that can be used by a medical examiner, forensic DNA testing may be necessary to establish identity. Autosomal and mitochondrial DNA testing can be helpful in determining the identity of unidentified human remains in missing-person and mass-disaster cases. In some cases, identification can be made using DNA profiles from relatives to establish relatedness or paternity.

8.1.4 Kinship

It might be necessary to establish kinship relationships when the parents of a person are not available for testing to establish paternity. In this case, DNA samples will be collected from other related individuals in order to establish a familial relationship. It should be noted that in any autosomal DNA case,

History and Standards of DNA Evidence

considerations must be made when parties are related because related individuals are expected to share more alleles.

When dealing with close relatives, such as parents and their children, full and half siblings, or cousins, calculations can be performed to determine the amount of shared genetic information. Likelihood ratios are used to weigh the proposition that the individuals being compared have a common ancestry against the proposition that they do not share a common ancestry. Paternity comparisons are based on the idea that a child receives one allele at every location from each parent, barring mutation. So then, a mother with genotype (8,12) and a father with genotype (11,14) may produce children with the following genotypes: 8,11; 8,14; 12,11; 12,14. This likelihood ratio calculation comparing hypotheses of relatedness came to be known as the Paternity Index (PI). With the PI, the numerator assumes paternity while the denominator assumes a random man is the father. The PI is calculated for each locus, then cross multiplied to give the combined PI. Generally, a PI of 100 is the minimum standard for an inclusion. If the PI is 100, that means that the alleged father has a 99 to 1 better chance of being the father than a randomly selected, unrelated male.

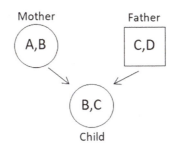

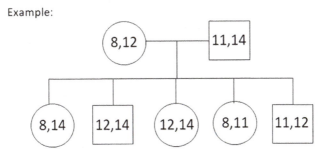

Figure 8.5 In Mendelian inheritance, the mother and father each pass one allele on to the child. An example pedigree is shown where the parents possess different alleles, which enables their children to have many different allele combinations.

160   Guide to the ABC Biology Exam

8.1.5 Databanking

The DNA Identification Act of 1994 established the Combined DNA Index System, or CODIS, which is the DNA database used in the United States. CODIS is maintained by the FBI for law enforcement purposes only. CODIS contains records of DNA profiles, both from crime scene evidence and known individuals. The CODIS database has 20 standard locations, or core loci, which are compared when profiles are entered. There are three levels to the CODIS database. The National DNA Index System, or NDIS, is maintained by the FBI and is an interstate database with strict requirements for entry. The State DNA Index System, or SDIS, is an intrastate database with relaxed standards for upload, but which only searches against profiles collected and uploaded in that state. The Local DNA Index System, or LDIS, has the most relaxed requirements for upload and very little oversight. Profiles uploaded at the local level are only searched against profiles uploaded by that laboratory.

Within the CODIS database, there are a number of indices. For criminal purposes, there is a convicted offender index, which contains DNA profiles from known persons who have been convicted of certain crimes. There is an arrestee index, which contains DNA profiles of individuals who have been arrested for a qualifying offense. And finally, there is a forensic index, which contains crime scene samples from evidence collected in relation to crimes. For other purposes, there is a missing persons index and a biological relatives of missing persons index to assist in identifying human remains.

Upon entry of a new DNA profile to CODIS, the alleles are added and searched against profiles already stored. If there is a match between a stored DNA profile from a crime scene and the new DNA profile of a recently arrested individual, this may indicate a link between the individual and the crime. Similarly, a new DNA profile obtained from a piece of crime scene evidence may match the DNA profile of a previously submitted crime scene evidence profile, which could connect the two crimes. DNA databases exist in other countries as well. The United Kingdom has one of the largest DNA databases in the world. The change from the previous 13 CODIS core loci to 20 loci in 2017 was due in part to the desire to cross compare profiles in international databases. DNA databases can be vitally important in cases where law enforcement has no suspects. In cases such as these, the database is invaluable for generating investigative leads.

8.2 Evolution of the Discipline

Forensic biology has grown to the discipline it is today through the development of three foundational applications used to associate an item of

History and Standards of DNA Evidence   161

evidence with a person of interest: antigen polymorphism, protein polymorphism, and DNA polymorphism. Each is an update on the prior method that allows for higher discriminatory power. The term "polymorphism" indicates that many forms of the same trait occur. For example, many forms of antigens occur and are expected to vary from person to person. Thus, the antigens found in a blood pool at a scene and the antigens found in the bloodstain on a suspect's jacket can provide a link if the antigens in both samples are identical. Likewise, many different forms of proteins occur in the body, and DNA has many variant regions that are unique to each individual. By studying the antigen, protein, or DNA profile similarities and differences, comparisons can be made between victims, suspects, and crime scene evidence. Each mechanism and its uses are explained in this section.

8.2.1 Antigen and Immunological Systems

Antigens are substances that reside on the surface of red blood cells and stimulate the body to produce antibodies against them. Immunogens are foreign substances capable of eliciting the antibody formation when introduced into a host. Immunogens usually consists of multiple epitopes, or binding sites, for antibodies to attach. Because of this, they are said to be multivalent. Haptens are small molecules that can also elicit an immune response. Antibody tests are used in the detection of drugs in blood and urine because the drugs are haptens that cause antibodies to bind to them.

Antibodies, also called immunoglobulins, are capable of binding specifically to an antigen. There are five main classes: IgG (which is the most abundant), IgA, IgM, IgD, and IgE. A typical antibody has two identical binding sites and is thus considered bivalent. The binding affinity and specificity of antibodies makes them useful reagents for serological testing, and there are two main types used: polyclonal and monocolonal.

Polyclonal antibodies are multivalent immunogens capable of eliciting a mixture of antibodies with diverse specificities. Polyclonal antibodies are produced by different B lymphocyte clones in response to the different epitopes of an immunogen. They are typically created by introducing a specific immunogen into a host animal. The animal will have an immune response to the immunogen and blood of the animal will be harvested and separated to isolate the antibodies. Rabbits, mice, birds, and horses are often used as host animals to produce polyclonal antibodies. The harvested antibodies are often referred to as antisera (because the antibodies are located in the serum of the blood) or antiglobulins.

Monoclonal antibodies are antibodies produced by a single clone of cells or cell line and consist of identical antibody molecules. Most often, monoclonal antibodies are derived from spleen cells that are harvested and inoculated with immunogens. A type of white blood cell, called a B lymphocyte, fuses with myeloma cells to create hybridoma cells. They are produced when an animal is injected with an antigen that provokes an immune response. Different types of white blood cells respond, among them B cells. The B cell produces antibodies that bind to the injected antigen. The new antibodies are harvested and the isolated B cells are fused with myeloma (cancer) cells to produce a hybrid cell line called a hybridoma. Hybridoma cells proliferate indefinitely in culture, meaning they are immortal.

The strength of the antigen–antibody binding is mediated by the interaction between the epitope of the antigen and the binding site of the antibody. Strong binding occurs only if the shape of the epitope fits the binding site. This depends on the affinity and avidity of the interaction. Affinity is the energy of interaction between a single epitope on an antigen and a single binding site on a corresponding antibody. The strength of this interaction depends on the specificity of the antibody for the antigen. Avidity is the overall strength of binding between antibody and antigen, considering all binding sites. Antigens are usually multivalent and antibodies are

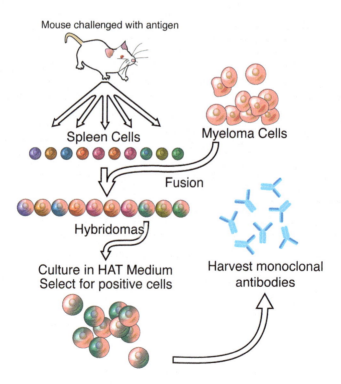

Figure 8.6 A general representation of the hybridoma method used to produce monoclonal antibodies.

History and Standards of DNA Evidence

usually bivalent. The avidity reflects the overall stability of an antigen–antibody complex.

There are three types of antigen and antibody binding reactions. There are *primary* binding reactions, where a single epitope of an antigen binds to a single binding site of an antibody. In this reaction, the antibody binds directly to the antigen rather than to another antibody. Tests employing this type of binding reaction are the most sensitive. *Secondary* binding reactions are those in which the antibodies bind to primary antibodies and not directly to the antigen. Some secondary binding reactions that are used in forensic science are precipitation reactions, agglutination reactions, and complement fixation. These secondary binding assays are less sensitive but easier to perform. There can also be tertiary binding, but tertiary assays are not commonly employed in a forensic setting.

The discovery that blood type is inherited and that the frequencies of each type vary in the population led to the use of an antigen polymorphic marker in forensic human identification cases. This was a step forward from blood group systems in forensic serology. The addition of the antigen polymorphism marker made results more discriminatory and lowered the probability of coincidental matches between persons unrelated. After A-B-O blood typing, antigen polymorphisms were the first method for linking a person of interest with forensic biology evidence from a crime scene, even before protein or DNA profiling.

Today, antigen and immunological assays have been able to expand even further. Antigens beyond A, B, and D can be identified through immunoassay techniques. These techniques are frequently used for serological screening in forensic biology. Currently, they are also employed in other areas of forensic science, such as toxicology. For example, immunoassay cards based on antigen–antibody response are often used to detect drugs in blood and urine. To begin an immunoassay, an antibody is needed to react with the antigen in question. Most often, these antibodies are produced in a process using animals, such as rabbits or mice. The target antigen is injected into the animal, where it stimulates an immune response. Serum is then drawn from the animal that will contain antibodies of interest. When the evidence sample is introduced to the antibodies from the serum, an agglutination reaction will occur if the sample contains the antigen of interest. This method of testing has been simplified into easy to use cards, which are similar in their chemistries and appearance to a pregnancy test. The ease and relatively low cost of these assays allows for high-volume specimen processing. Commercially prepared serum taken from animals injected with a variety of drugs is designed to react with many drugs at once. Some immunoassays are presumptive in nature, but others, such as the ABAcard Hematrace® assay, are confirmatory.

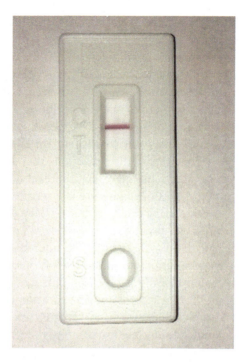

Figure 8.7 Immunochromatographic assay for detecting human blood. This assay functions based on antigen–antibody interactions.

8.2.2 Protein and Enzyme Polymorphisms

Protein polymorphism was introduced next to further expand the discrimination power of A-B-O typing and antigen polymorphism. The combination of blood groups and protein polymorphisms significantly lowered the probability of a match between two unrelated people. This laid the foundation for the STR profile we know today, and it is based on the same principle. For each protein polymorphism that was examined, the blood profile became more and more unique. Just as, with each locus examined on a person's DNA, the DNA profile becomes more and more unique. It is estimated that 20–30% of proteins in humans are polymorphic, and protein profiles can be generated based on the types of protein polymorphisms present in each individual. These variations allow suspects to be more easily eliminated.

Enzymes are proteins that exist to catalyze reactions in the body. Red blood cells contain isoenzymes, which are various versions of an enzyme that all catalyze the same reaction but have different amino acid sequences. Isoenzymes are another factor that can further differentiate

History and Standards of DNA Evidence

individuals. An example of an erythrocyte polymorphism that was historically important in forensics is phosphoglucomutase, or PGM. It was one of the first protein markers examined forensically, and it was also present in semen stains, which made it forensically relevant for sexual assault investigations.

Another forensically relevant protein is hemoglobin (Hb). More than 200 Hb variants have been identified, which can be useful in characterizing blood. For example, hemoglobin S is found primarily in people with Hispanic or African heritage. A heterozygous individual with only one copy of the Hb S gene is said to carry the sickle cell trait, and this variant can be identified using electrophoresis. Lastly, blood serum protein polymorphisms, or immunoglobulin proteins, can be detected in bloodstains and have the ability to distinguish individuals.

Protein and enzyme polymorphisms can be identified through electrophoresis. This process separates the proteins based on their molecular weights and charges. This is usually done in one of two matrices: papers, such as cellulose acetate, or gels, such as starch, agar, agarose, or polyacrylamide. If a certain protein is unable to be separated well by electrophoresis in its native form, it can be denatured in a process called denaturing protein electrophoresis. A third method of protein separation is by use of their isoelectric points (pI) through isoelectric focusing (IEF). A protein's pI is the pH at which its net charge is zero. IEF creates a pH gradient in a gel between two electrodes. The protein enters the gel and migrates until it reaches the pH equal to its pI value.

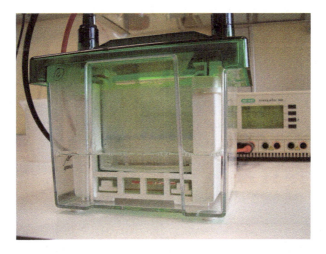

Figure 8.8 Protein polyacrylamide gel electrophoresis apparatus.

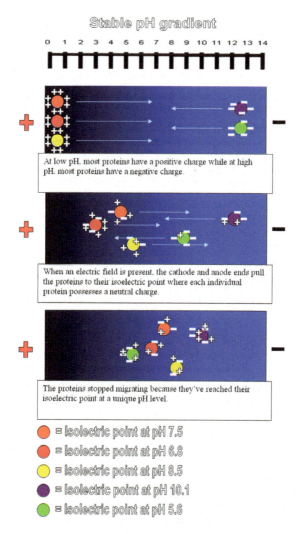

Figure 8.9 IEF electrophoresis explained.

Figure 8.10 RFLP fragments. The splicing of DNA from two individuals produces different-sized fragments, which can be separated on a gel to show the differences in the genes between the individuals.

History and Standards of DNA Evidence

8.2.3 DNA Polymorphisms

8.2.3.1 RFLP

The culmination of prior antigen and protein profiling technologies led to the discovery of DNA polymorphisms. DNA "fingerprinting" was first used in a criminal investigation in the 1980s in the United Kingdom. The case of Colin Pitchfork, the Black Pad Killer, demonstrated the potential power of this new technology to assist in solving crime. DNA polymorphisms have the unique ability to not only include potential suspects, but also exclude innocent people from the search for the perpetrator. Many people don't know that in the Black Pad case, the police had a suspect in custody who was excluded as a source of the DNA on the victims before Colin Pitchfork was ever arrested. The very first application of forensic DNA to a criminal case exonerated an innocent person.

Interest in DNA polymorphisms was launched by the work of the Human Genome Project, which sequenced the entire nuclear human genome. It was the research scientists on this project who identified and first described the use of RFLP. Restriction fragment length polymorphism (RFLP) was the first method used in forensic DNA testing. RFLP requires double-stranded, intact DNA samples, which in itself is a limitation given that many samples from crime scenes are exposed to the elements and thus may be partially degraded. At a crime scene, having high-quality, high-quantity DNA is not extremely common. In RFLP, enzymes known as restriction endonucleases, or restriction enzymes, recognize certain DNA sequences and cleave the phosphodiester bonds in the DNA upon encountering such sequences. Each restriction endonuclease will cut the DNA at every location where it recognizes the special sequence of nucleotides. Hundreds of restriction endonucleases are used so the DNA can be cut at the areas of interest required for a particular assay.

Multiple locations on the DNA were studied by scientists to determine their suitability in distinguishing individuals. To qualify for forensic application, the areas where the polymorphisms were located first had to be identified in the genome. These locations had to be inherited independently to be selected. Once the locations were selected, endonucleases had to be designed to cut the areas of interest. These sections of DNA, called variable number of tandem repeats (VNTR's), contained core repeats between 10 and 100 bases long. The resulting cut DNA fragments vary by length and are separated by gel electrophoresis in an agarose gel.

After being separated in the gel, the DNA is denatured and transferred to a matrix where it is immobilized and hybridized with a labeled probe in a process known as Southern transfer and hybridization. The pieces of DNA that complement the sequence of the labeled probe can be detected using either autoradiography or chemiluminescence.

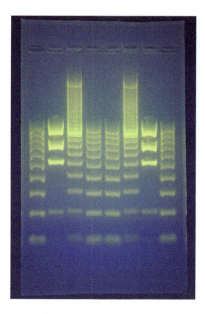

Figure 8.11 Chemiluminescent detection of DNA after separation by electrophoresis.

8.2.3.2 PCR, qPCR, RT-PCR

8.2.3.2.1 Polymerase Chain Reaction In the mid-1980s, a research scientist named Kary Mullis, who worked for Cetus Corporation, developed a new technique for copying pieces of DNA called polymerase chain reaction (PCR). Using PCR, a small quantity of DNA can be amplified exponentially, which makes DNA analysis a possibility even when minute amounts of DNA are found on a crime scene sample. With PCR, scientists moved away from the previous RFLP and VNTR loci, which required fairly long sequences of high quality DNA. Instead, they embraced microsatellite regions called short tandem repeats, or STRs, because PCR could make billions of copies of the regions of interest and it was no longer necessary to use large pieces of DNA. Research was performed to identify locations that were still highly variable, but with much shorter pieces of code. STR loci were characterized through research, and scientists moved away from locations with 10–100 base pairs in the core repeat towards areas consisting of four or five base pair codes. These STR segments were less susceptible to environmental insults, like degradation, and could be more efficiently copied.

Once the STR loci had been identified and characterized by researchers, most labs began by working with a small set of loci, often four or five. While this set of loci was helpful in excluding individuals, it could still produce false associations between unrelated individuals and evidence. As more STRs were researched and characterized, more loci were added to standard testing kits. With the addition of each new locus, the discrimination power of the STR PCR process increased.

History and Standards of DNA Evidence

The FBI initially standardized 13 STR loci to be the core loci that had to be tested for a sample to be stored in the CODIS database. In January 2017, they expanded the original 13 core loci to 20 in order to allow for more discrimination power and international database compatibility.

PCR exponentially amplifies specific sequences of DNA to yield an amplified product known as an amplicon. The more copies that are made in the

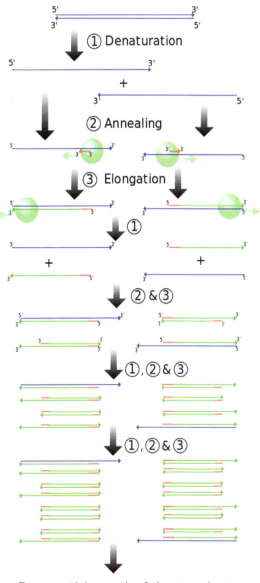

Figure 8.12 The steps of the PCR process. Each of these steps takes place in a cycle that is repeated 28–32 times in a normal PCR amplification.

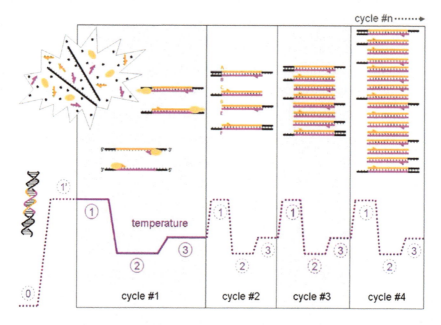

Figure 8.13 A representation of the PCR cycle, showing exponential amplification along with the relative temperature levels of each phase.

PCR process, the greater the chance the analyst will obtain some result. PCR is a very sensitive technique capable of amplifying extremely small amounts of DNA, which helps to overcome some of the limitations of the technology that involve small sample size.

Each cycle of PCR consists of three phases: exponential, linear, and plateau. The exponential phase is where all of the DNA targeted is doubled each cycle. During the exponential phase, the reaction efficiency is 100%, as all PCR reagents are still in excess to the amplicon. The more DNA that was input into the system, the more copies that will be made. In the linear phase, PCR reagents begin to get used up, and the reaction efficiency drops below 100%. During the plateau phase, no more copies of DNA are made because the reagents and polymerase have been completely exhausted. In this phase, the reaction efficiency is 0%.

8.2.3.2.2 qPCR and Real-time PCR Since the process of PCR is extremely sensitive, it is necessary to know how many DNA copies are being created throughout the replication process. In forensic DNA, an informal "Goldilocks Principle" is applied. If your PCR input amount is too high, the resulting DNA profile will be rife with artifact, such that it may be difficult to interpret. If your PCR input amount is too little, incomplete results could be obtained that may limit the evidentiary value of the testing. To ensure the proper amount is inputted into the system, an estimation of how much

History and Standards of DNA Evidence

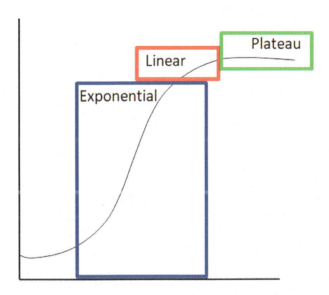

Figure 8.14 A graph depicting the PCR amplification curve.

DNA is present in the crime scene sample must occur. This is done through quantitative PCR. Quantitative PCR (qPCR) assays include end-point PCR and real-time PCR. End-point PCR measures how much DNA product is present at the end of the amplification process, after the plateau phase. The DNA concentration of an amplified product should directly correlate to the amount of DNA initially tested, assuming that the amount of DNA present doubles every cycle.

Real-time PCR quantifies the DNA amplified during the exponential phase of PCR. This is accomplished by fluorescent dyes that can either intercalate into double stranded DNA, such as SYBR Green, or with dyes that specifically anneal to target areas on the DNA, such as Taqman. The number of cycles it takes for the DNA signal in the sample to cross a defined threshold is known as the cycle threshold. This value can be used to determine how much DNA is in the original sample. This method is slightly more precise than end-point PCR since it is not affected by any variations in PCR conditions. Real-time PCR is the most widely used form of quantitative PCR since it can monitor the amplification process throughout each cycle.

8.2.3.2.3 The PCR Cycle and Its Reaction Components The reaction components utilized in PCR include thermostable DNA polymerases, primers, and DNA nucleotides. If utilizing real-time QPCR, a dye or fluorescent probe will also be included. Taq polymerase is the thermostable DNA polymerase most widely used in PCR for its efficiency and ability to recreate long DNA fragments with precision.

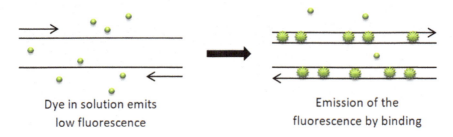

Figure 8.15 The mechanism by which fluorescence occurs in end-point PCR.

Taq polymerase from Applied Biosystems, AmpliTaq Gold DNA polymerase, is the most frequently used in forensics. AmpliTaq Gold DNA polymerase is altered to be inactive until it is at a pH below 7. The buffer in PCR changes pH with temperature: as temperature increases, pH decreases. Thus, to activate AmpliTaq Gold, the system is heated to 95°C before the first cycle begins, in a process known as hot start PCR. This heat causes the DNA to denature, which limits the possible formation of non-specific PCR products. PCR primers are the second component necessary for the PCR process. Primers flank the target region of the DNA template around an area of interest and enable that section to be copied. These primers are oligonucleotides, or short DNA segments, which complement the sequences around the target; one forward and one reverse primer anneal to each strand of the denatured DNA. Primers are designed using a computer software program based on the PCR requirements and they must complement the sequences around the target sequence perfectly. They should be around 15–25 base pairs in length, with 40–60% of the bases being G or C.

Further, a primer should not contain any complementary sequences within its own strand or with its partner strand or else it may form a hairpin

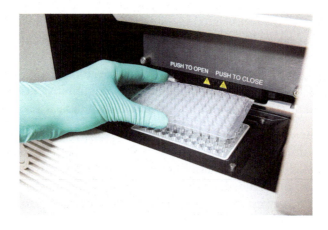

Figure 8.16 A scientist loading a sample into a real-time PCR instrument.

History and Standards of DNA Evidence

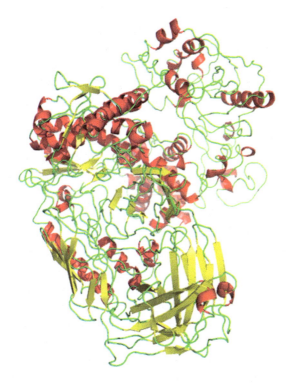

Figure 8.17 Taq polymerase is the most important enzyme in PCR, which attaches the nucleotides to the parent strand.

structure with itself or anneal to the other primer rather than to the DNA. Amplified primers are called primer dimers, and they can interfere with the amplification of the target DNA sequence. Other components that are essential for successful PCR assays include template DNA with a linear target sequence in the case of nuclear DNA, or circular target sequence for mitochondrial DNA. These sequences can either be single- or double-stranded. Additionally, free nucleotides are present in the PCR solution to be used by the Taq polymerase to extend the DNA sequence.

Each phase of the PCR cycle is temperature specific. The first stage, denaturation, happens at high temperatures around 94–95°C. Temperature is then lowered for the annealing of the primers and template. The annealing temperature depends on the primer being used, but is typically around 50–60°C. Lastly, extension occurs at 72°C, which is the optimal temperature for the DNA polymerase to replicate the DNA.

These three steps (denature, anneal, extend) are repeated as many times as is necessary to create a DNA sample large enough for profile detection. PCR usually takes place in a thermal cycler so temperature can be changed and monitored exactly. Lastly, positive and negative controls should be used to show that PCR components all function properly during each cycle.

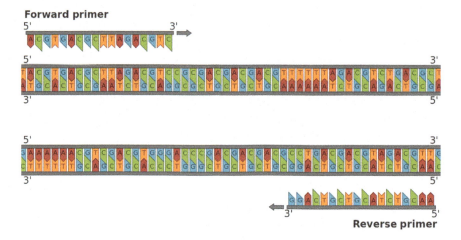

Figure 8.18 PCR primers are short sequences of single-stranded DNA that are the same as one side of the parent strand. These sequences flank the sequence of interest, usually regions containing SNPs or other variable regions.

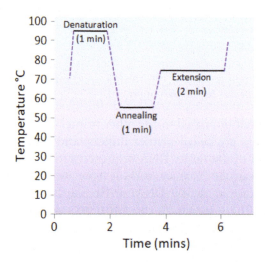

Figure 8.19 This graph shows the temperatures at which each cycle of PCR takes place.

8.2.3.2.4 Potential Factors Affecting PCR Contamination, degradation, and inhibition must all be monitored to ensure optimal PCR performance. It is possible for a sample to be contaminated by other samples or by the person performing the test. One important step for preventing sample to sample contamination of a DNA sample is preparing and processing the pre- and post-PCR samples in different places and at different times. Supplies, reagents, and equipment should be treated similarly, one set for

History and Standards of DNA Evidence

pre-amplification analysis and one set for post-amplification analysis. To prevent contamination from an analyst to the sample, the analyst should wear a laboratory coat, disposable gloves, and a facial mask at all times when processing samples in the lab.

If a sample becomes contaminated, this can be seen through the use of controls. A reagent blank contains the reagents used at every stage of the analysis but should not contain DNA. Blank controls are run alongside the evidence samples. An amplification negative control will also be run alongside the samples, and this contains all amplification reagents and no DNA.

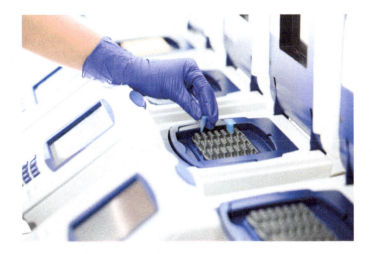

Figure 8.20 Thermal cycler used in PCR amplification.

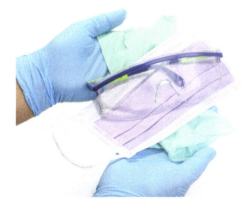

Figure 8.21 Proper personal protective equipment for a DNA analyst: laboratory coat, disposable gloves, facial mask, eye goggles, and hair cap.

If contamination is found in the reagent blank, an investigation will begin to find out in which stage the contamination may have been introduced. Contamination found in an amplification negative control but not in the reagent blank indicates that contamination occurred at some point during the amplification step. All labs should have a repository of analysts' DNA profiles available in the lab for comparison to determine if contamination may have occurred.

Positive controls are also utilized in the DNA testing and PCR process. An amplification positive control contains known DNA, such as DNA Control 007, that is included in amplification kits such as Applied Biosystems GlobalFiler kit. The positive control is used to demonstrate that the reagents and instruments used during the amplification process are working properly. If the proper known DNA profile can be generated, then everything is working as it should. If not, an investigation into the problem will commence.

Template quality may complicate the PCR testing process. If DNA is degraded, it may affect amplification efficiency (especially at higher molecular weight loci) or prevent amplification of the sample completely. Degradation is commonly observed in samples that are old or have been stored in suboptimal conditions, such as high humidity or direct sunlight, for some time. Optimal conditions for high-quality DNA require it to be stored in a cool, dry place. Generally, most samples at crime scenes have not been maintained in an ideal environment. If samples have been collected from the scene, they should always be allowed to fully dry prior to packaging, and be stored in cardboard boxes or paper bags that will not trap moisture. Failure to do so could cause mold or bacteria to grow on the sample, which may result in PCR failure at some or all loci.

Inhibitors can cause PCR amplification failure, likely through interfering with the proper function of Taq polymerase. Inhibitors in routine casework can be from heme on items bearing blood, indigo dyes from denim, tannin from leather, or melanin from hair. Also, bacteria from soil and samples that are extremely "dirty" and have not been sufficiently filtered after extraction can show inhibition. Inhibitors can be detected at the quantitation step using an internal positive control. All modern real-time PCR kits used for quantitation have an indicator for inhibition.

DNA extraction and preparation for PCR must be performed properly; if inhibitors remain in the sample, they can produce suboptimal results or no results at all. Filtration and washing may be used to remove any inhibitors not eliminated in the extraction process; inhibitors may also be eliminated by increasing DNA polymerase quantities or adding bovine serum albumin (BSA) into the reaction. Today's robotic extraction systems do a superior job of removing inhibition from samples as well.

Table 8.1   List of PCR Amplification Kits that Demonstrates whether PCR Inhibition Is Detected along with Limits of Detection

Kit/Assay	Detection Method	Limit of Detection	Volume of DNA used	Human/Primate Specific	PCR Inhibition Detected	References
Quantiblot Kit	Hybridization	~150pg	5μL	Y	N	Walsh et al. (1992)[1]
PicoGreen Assay	Intercalating Dye Fluorescence	250pg	10μL	N	N	Hopwood et al. (1997)[2]
AluQuant Kit	Pyrophosphorylation and luciferase light production	100pg	1–10μL	Y	N	Mandrekar et al. (2001)[3]
BodeQuant	End-point PCR	~100pg	1–10μL	Y	Y	Fox et al. (2003)[4]
TQS-TH01	End-point PCR	~100pg	1–10μL	Y	Y	Nicklas and Buel (2003a)[5]
Quantifiler Kit	End-point PCR	20pg	2μL	Y	Y	Applied Biosystems (2003)[6]
Alu Assay	End-point PCR	1pg	1–10μL	Y	Y	Nicklas and Buel (2003c)[7]
CFS TH01 Assay	End-point PCR	20pg	1–10μL	Y	Y	Richard et al. (2003)[8]
RB1 and mtDNA multiplex	End-point PCR	20pg	1–10μL	Y	Y	Andreasson et al. (2002)[9]

8.2.3.2.5 Reverse Transcriptase PCR (RT-PCR)

Recall, from the process of transcription, that DNA serves as a template to make a complementary RNA strand. Likewise, it is possible to use RNA as a template for complementary DNA (cDNA) through a process known as reverse transcription. An enzyme appropriately named reverse transcriptase performs this process. This enzyme is similar in structure and function to DNA polymerase, but genetically engineered. Reverse transcriptase PCR (RT-PCR) is a mechanism for detecting small quantities of mRNA to synthesize a single strand of cDNA from the template mRNA, and then amplifying the cDNA through PCR for analysis.

Like in regular PCR, a primer that can attach to the mRNA template is required for RT-PCR. This primer can be either for RNA or DNA, though DNA primers are more efficient. Gene-specific primers anneal to specific mRNA sequences to convert a particular gene sequence into cDNA. Universal primers anneal to any mRNA sequence so all mRNA is converted to cDNA. Universal primers include oligo (dT) and random hexamer primers. Oligo (dT) primers anneal to the 3' poly-A tails on eukaryotic mRNA. Random hexamer primers nonspecifically anneal at many sites on any RNA strand, even ribosomal RNA.

RT-PCR can happen in one step or two steps. In one-step RT-PCR, the reverse transcription and PCR occur in the same tube. Gene-specific primers are the only primers that can perform both reverse transcription and PCR. In this one-step procedure, the risk of contamination is reduced, and the reaction is usefully simplified when handling a large number of samples. Two-step RT-PCR separates the process of reverse transcription from PCR and carries out both in separate tubes. Two-step RT-PCR converts all of the RNA into cDNA, and thus it is useful for analyzing multiple mRNA strands in a single sample. Two-step RT-PCR is more commonly used for forensic identification.

8.2.3.3 *Genetic Markers*

Certain sequences on DNA are the same from person to person, but it is the sequences that are unique to each individual that pique the interest of the forensic scientist. DNA polymorphisms can differ in sequence of nucleotides (sequence polymorphism) or number of tandem repeats (length polymorphism). The majority of DNA polymorphisms are only differences of one nucleotide, known as single nucleotide polymorphisms (SNPs). More than one million SNPs have been found in the human genome. However, longer polymorphisms are utilized in DNA profiling as well.

Tandem repeats are units of repeated nucleotides located side by side in the DNA sequence. Satellite DNA is a type of tandem repeat, along with minisatellites and microsatellites, each classifying a different length of a stretch of tandem repeats. Variable Number Tandem Repeats (VNTRs) are minisatellites with a high percentage of A and T nucleotides ranging in length from several to hundreds of base pairs. The number of repeated nucleotide segments varies

History and Standards of DNA Evidence

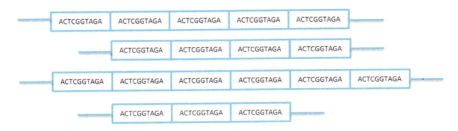

Figure 8.22 VNTRs are sequences of nucleotides that repeat a number of times in DNA. Then the number of repeats varies from person to person.

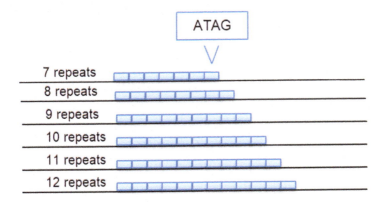

Figure 8.23 STRs are shorter than VNTRs in length but are similar in that the number of repeats varies from person to person.

from person to person along certain VNTR loci, which leads to possibilities for the separation of suspects. VNTR profiling uses RFLP techniques.

Short Tandem Repeats (STRs) are microsatellites, and are only 2-6 base pairs in length. Two unrelated individuals have never been documented to have the same STR or VNTR polymorphisms. There are three main types of tandem repeats found at polymorphic loci used in forensic DNA analysis. Simple sequence repeats consist of one repeating sequence. Compound repeats comprise two or more adjacent simple repeats, and complex repeats may contain several repeat blocks of variable unit length as well as variable intervening sequences.[10] Loci that are single sequence repeats are TPOX, CSF1PO, D5S818, D13S317, D16S539. Some loci have simple sequence repeats with non-consensus alleles (e.g., 9.3) like TH01, D18S51, D7S820. Other loci are comprised of compound repeats with non-consensus alleles (VWA, FGA, D3S1358, D8S1179) and the locus D21S11 is comprised of complex repeats.

Another form of genetic markers are mobile elements, randomly dispersed repeats that can change their location through a process called transposition. The two human types of mobile elements are DNA transposons and

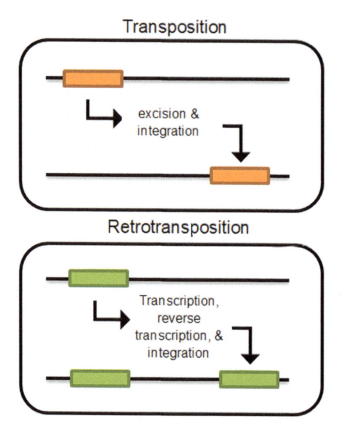

Figure 8.24 Transposon and retrotransposon action.

retrotransposons. Through a "cut-and-paste" movement, DNA transposons are removed from one region and inserted into another place on the genome. Through a "copy-and-paste" movement, retrotransposons duplicate themselves and insert the copies in another area of the genome.

Over the last 20 years, a number of STRs have been explored and characterized for forensic use. STR loci were selected for use in human identification applications based on the following characteristics[11,12]:

- High discriminating power, usually >0.9, with observed heterozygosity >70%.
- Separate chromosomal locations to ensure that closely linked loci are not chosen.
- Robustness and reproducibility of results when multiplexed with other markers.
- Low stutter characteristics.

History and Standards of DNA Evidence

- Low mutation rate.
- Predicted length of alleles that fall in the range of 90–500 bp with smaller sizes better suited for analysis of degraded DNA samples.

The original 13 core STR loci were selected for standardization of the CODIS national DNA database.[13,14] The original 13 CODIS core loci are CSF1PO, FGA, TH01, TPOX, VWA, D3S1358, D5S818, D7S820, D8S1179, D13S317, D16S539, D18S51, and D21S11. Of these, the three most polymorphic markers are FGA, D18S51, and D21S11, while TPOX shows the least variation between individuals.

In 2015, the FBI announced they would be adding an additional seven loci—D1S1656, D2S441, D2S1338, D10S1248, D12S391, D19S433, and D22S1045—along with the original 13 loci. The full list of all 20 CODIS core loci is as follows:

- CSF1PO
- D3S1358
- D5S818
- D7S820
- D8S1179
- D13S317
- D16S539
- D18S51
- D21S11
- FGA
- TH01
- TPOX
- vWA
- D1S1656 (effective January 1, 2017)
- D2S441 (effective January 1, 2017)
- D2S1338 (effective January 1, 2017)
- D10S1248 (effective January 1, 2017)
- D12S391 (effective January 1, 2017)
- D19S433 (effective January 1, 2017)
- D22S1045 (effective January 1, 2017)

8.2.3.4 Automation

With the large amount of evidence that needs processing in the modern forensic lab, procedures that can be automated and used for reliable, reproducible results are exceedingly beneficial to all involved. Because of this reality, many forensic labs specifically select processes based on their amenability to automation.

In Chapter 10, we will discuss all of the different methods for DNA extraction. Silica-based extraction methods are quickly gaining popularity because many robotic systems employ them. There are two main types of laboratory automation in use in the modern forensic biology lab: liquid handling and interpretation.

The use of liquid handling robots to perform DNA extraction, PCR setup, and PCR amplification analysis steps increases lab productivity while reducing the amount of human error that comes with manual manipulation. Liquid handling robots allow laboratories to increase their throughput capacity while reducing the workload on their laboratory personnel.

Expert systems and probabilistic genotyping software are two methods for interpreting DNA profile data that help to ease the workload on the modern forensic DNA workforce. For many high-throughput laboratories, data assessment and interpretation comprise a significant time investment. For mixed samples, the time spent interpreting a DNA profile can be exponentially greater than the time involved in developing the profile at the lab. Acknowledging this, software has been designed and implemented to replace the manual human interpretation. Expert systems translate the electropherogram signal into a genotype compatible with a database. This type of software is generally used for known exemplars and database samples containing single-source DNA profiles.

Probabilistic genotyping programs, such as TrueAllele and STRmix, are software programs that analyze DNA profile data to aid in the deconvolution of complex mixed DNA samples. The interpretation of mixed DNA profiles has gotten more and more difficult as the sensitivity of the technology has increased. In the past, analysts have used a "binary" approach to interpretation, where inferred genotypes are either included or excluded from the mixture. To do this, they employed a series of parameters such as a stochastic threshold, sister allele (heterozygote) balance, mixture ratios, and stutter ratios.

There are two main types of probabilistic genotyping software programs: semi-continuous and fully continuous. The semi-continuous method can incorporate a probability for dropout, $Pr(D)$, and a probability for drop-in, $Pr(C)$. Semi-continuous methods do not use peak heights when generating possible genotype sets and do not model artifacts such as stutter. LR Mix Studio is an example of an open-source semi-continuous probabilistic genotyping software program.

Fully continuous programs, by contrast, model allelic and stutter peaks as well as quantitative information from the DNA profile. They calculate the probability of the peak heights given all the possible genotype combinations for the individual contributors. These continuous methods make

History and Standards of DNA Evidence

assumptions about the underlying behavior of peak heights to evaluate the probability that two peaks would be paired together. They use more information from within the profile and reduce the requirement for the subjective manual interpretation. TrueAllele, developed by Cybergenetics, and STRmix, developed by ESR, are examples of fully continuous probabilistic genotyping software programs.

8.3 Accepted Standards and Practices

8.3.1 Quality Assurance Standards for Forensic DNA Laboratories

As discussed in previous chapters, strong quality assurance is a requirement for forensic laboratories to ensure that certain processes and services meet standards for test integrity. The FBI publishes Quality Assurance Standards (QAS), to which all public crime laboratories who wish to utilize the CODIS national DNA database must adhere. In addition to the FBI QAS, each accrediting body will have quality assurance standards they impose on laboratories maintaining accreditation in their program. Moreover, there are guidelines published by the Scientific Working Group on DNA Analysis Methods (SWGDAM) that outline definitions and rules for quality assurance in laboratories. For example, SWGDAM recommends that each laboratory document the following elements, as listed in the SWGDAM document from September 1, 2011: goals and objectives, organization and management, personnel, facilities, evidence control, validation, analytical procedures, equipment calibration and maintenance, reports, review, proficiency testing, corrective action, audits, safety, and outsourcing. The quality system for each DNA lab is annually reviewed and audited.

8.3.2 Quality Assurance Standards for DNA Databasing Laboratories

Slightly different to forensic DNA testing laboratory standards are the standards for DNA databasing laboratories. These labs perform DNA analysis for the purpose of entering the DNA profile into a database (primarily, CODIS). The FBI has a specific set of standards for labs of this type. SWGDAM also covers the quality assurance definitions and guidelines for databasing laboratories. Only slight variations exist between this document and the guidelines for DNA testing laboratories.

8.3.3 SWGDAM

The Scientific Working Group on DNA Anlaysis Methods, or SWGDAM, has published guidelines on the best practices for quality assurance for the forensic DNA community. SWGDAM was established by the FBI in 1988 and took over the provision and monitoring of guidelines in 2000. It covers all DNA laboratories in the US and Canada. Its members are currently working DNA analysts at the FBI and in public crime laboratories across the country. This group presents guidelines to the FBI director and makes recommendations for the FBI QAS. The FBI holds meetings with SWGDAM to receive input from the forensic community, discuss any laboratory methods that may need improvement, and share protocols for testing. Their guidelines are revised often as the field of forensic DNA advances.

8.3.4 CODIS

The Combined DNA Index System, CODIS, as introduced earlier in this chapter, is a database created by the FBI to store and search DNA profiles. There are three levels of the CODIS database: local, state, and national. There are sections of the database for missing persons and unidentified remains, convicted offenders and arrestees, and of course, crime scene evidence profiles. Any lab wishing to participate in CODIS must follow the FBI QAS and be accredited by a recognized entity that audits them against those standards.

8.3.5 Validation for Introduction of New Technologies

In Chapter 2, we discussed the concept of validation in detail. Since the field of forensics is constantly expanding and improving, it is important that any new technology that will be used on casework samples is thoroughly tested. Validation involves testing an instrument or methodology to discover its capabilities and limitations; it is a discovery process to determine the parameters in which the thing being tested performs optimally, accurately, and efficiently.

Two types of laboratory validations exist to modify forensic DNA analysis methods: developmental and internal validations. A developmental validation according to SWGDAM's Validation Guidelines for DNA Analysis Methods "is the acquisition of test data and determination of conditions and limitations of a new or novel DNA methodology for use on forensic, database, known or casework reference samples." Precision and accuracy are vital in a developmental validation. Internal validation occurs inside a laboratory that wishes to employ the new method. SWGDAM defines an internal validation as "an accumulation of test data within the laboratory to demonstrate that established methods and procedures perform as expected in the laboratory."

Notes

1. Walsh, P.S., Varlaro, J. and Reynolds, R. (1992) *Nucleic Acids Research*, 20, 5061–5065.
2. Mandrekar, M.N., Erickson, A.M., Kopp, K., Krenke, B.E., Mandrekar, P.V., Nelson, R., Peterson, K., Shultz, J., Tereba, A. and Westphal, N. (2001) *Croatian Medical Journal*, 42, 336–339.
3. Mandrekar, M.N., Erickson, A.M., Kopp, K., Krenke, B.E., Mandrekar, P.V., Nelson, R., Peterson, K., Shultz, J., Tereba, A. and Westphal, N. (2001) *Croatian Medical Journal*, 42, 336–339.
4. Fox, J.C., Cave, C.A. and Schumm, J.W. (2003) *Biotechniques*, 34, 314–322.
5. Nicklas, J.A. and Buel, E. (2003a) *Journal of Forensic Sciences*, 48, 282–291.
6. Applied Biosystems (2003) *Quantifiler™ Human DNA Quantification Kit and Quantifiler™ Y Human Male DNA Quantification Kit User's Manual.* Foster City, California: Applied Biosystems.
7. Nicklas, J.A. and Buel, E. (2003c) *Journal of Forensic Sciences*, 48, 936–944.
8. Richard, M.L., Frappier, R.H. and Newman, J.C. (2003) *Journal of Forensic Sciences*, 48, 1041–1046.
9. Andreasson, H., Gyllensten, U. and Allen, M. (2002) *Biotechniques*, 33, 402–411.
10. Urquhart, A., Kimpton, C.P., Downes, T.J. and Gill, P. (1994). Variation in Short Tandem Repeat sequences —a survey of twelve microsatellite loci for use as forensic identification markers. *International Journal of Legal Medicine*, 107, 13–20.
11. Gill, P., Urquhart, A., Millican, E. S., Oldroyd, N. J., Watson, S., Sparkes, R. and Kimpton, C. P. (1996) *International Journal of Legal Medicine*, 109, 14–22.
12. Carracedo, A. and Lareu, M.V. (1998) *Proceedings from the Ninth International Symposium on Human Identification*, pp. 89–107. Madison, Wisconsin: Promega Corporation.
13. Budowle, B., Moretti, T.R., Niezgoda, S.J. and Brown, B.L. (1998) *Proceedings of the Second European Symposium on Human Identification*, pp. 73–88. Madison, Wisconsin: Promega Corporation.
14. Budowle, B. and Moretti, T.R. (1998) *Proceedings of the Ninth International Symposium on Human Identification*, pp. 64–73. Madison, Wisconsin: Promega Corporation.

Bibliography

Forensic DNA typing, by John Butler (2005).

Li, R. C. (2015). *Forensic biology* (2nd ed.). Boca Raton: CRC Press (Ch 3, 5, 6, 7, 18, 19, 26).

Saferstein, R. (2018). *Criminalistics: an introduction to forensic science* (12th ed.). Boston: Pearson (Ch 1, 3, 14).

Shelton, D. (2011). *The 'CSI Effect': Does It Really Exist.* National Institute of Justice. https://www.rvrhs.com/ourpages/auto/2011/4/8/38856516/CSI%20Effect.pdf

SWGDAM. (September 1, 2011). Quality Assurance Standards for DNA Databasing Laboratories. Retrieved June 2, 2017, from https://www.swgdam.org/publications#!

SWGDAM. (September 1, 2011). Quality Assurance Standards for Forensic DNA Testing Laboratories. Retrieved June 2, 2017, from https://www.swgdam.org/publications#!

DNA Testing Process Part 1

9.1 Casework Documentation and Reporting (Results and Conclusions)

9.1.1 Case Management

In each forensic DNA testing lab, there will inevitably be someone that must organize and prioritize the cases for testing. The laboratory will likely develop policies regarding case and sample priority. Because DNA labs could never test every item of evidence collected from a crime scene, it's the duty of the laboratory to create a case management policy to set reasonable limitations on the number of samples that can be submitted or tested. This is often dependent on case type. Individuals in the lab would also be available to help investigators select samples that should be tested first depending on the likelihood that the samples will yield results.

Case management is an important step in the DNA testing process, because it helps streamline the workflow in the DNA lab for maximum efficiency. Cases are organized to ensure samples that need to be tested are completed in time for upcoming trials, or to meet any deadlines imposed by law. At many high-level labs, the case management unit or an evidence coordinator is also charged with redacting files to remove any investigative information that might be potentially biasing to the DNA analyst. In a process called "linear sequential unmasking," the case management personnel remove any references to confessions or other investigative theories that have no bearing on the downstream DNA analysis. In some cases, DNA analysts receive heavily redacted case scenarios or incident reports that only contain sanitized information necessary for analysis.

9.1.2 Evaluate Requests for Analysis to Determine Appropriate Evidence Screening and Comparisons to Develop the Most Useful Information

Upon receiving a piece of evidence for DNA testing, a DNA analyst must assess what case information is provided to ensure it will supply the type and kind of information that the investigator is seeking. If a sexual assault is alleged, serological tests for body fluids should be conducted. The analyst must ensure that the appropriate screening tests and examinations are executed, but they must also ensure that the samples selected to be tested are the most probative samples in the case. For example, if a sexual assault of a child occurs in a home with multiple male relatives of the same paternal lineage, samples would not be suitable for Y-STR testing because the results would not be able to differentiate between related individuals.

Also, it is the responsibility of each analyst to perform testing without disrupting or contaminating the evidence needed by the other units. If an item needs testing in both fingerprints and DNA, the analyst must ensure that the least destructive analyses are performed first. Analysts must think critically about what samples are selected for testing given their expertise and keeping in mind that investigators and case managers may not have in-depth knowledge.

Figure 9.1 This item of evidence could contain fingerprint evidence as well as biological evidence, so caution must be taken during processing to preserve both types of evidence.

9.1.3 Establishing Case Record

Beginning at the crime scene and continuing until destruction of evidence, every person who takes possession of an item of evidence and every place where that evidence is stored is documented in the chain of custody. Along with the chain of custody documents, paperwork is created at every stage of the forensic biology process, detailing the analyses and results generated. The culmination of this documentation is known as the case record. Once the evidence has been processed, a copy of the case record is kept in document storage in the lab or retained electronically.

The National Commission on Forensic Science made formal recommendations to the Department of Justice regarding what should be contained in the case record. Those recommendations are as follows:

The National Commission on Forensic Science (NCFS) recommends that the Attorney General take the following action:

DNA Testing Process Part 1 189

Figure 9.2 A case file storage room at a crime lab.

- *Department of Justice Forensic Science Service Providers (FSSPs) develop and maintain written policies for documenting the examination, testing, and interpretation of evidence and for reporting results, interpretations, and conclusions that are consistent with the following requirements:*
  1. *Records should be created contemporaneous with the examination of evidence and the technical review that, along with the FSSPs quality management system documents relating to the forensic work performed, would allow another analyst or scientist, with proper training and experience, to understand and evaluate all the work performed and independently analyze and interpret the data and draw conclusions.*
  2. *Providing all of the documentation encompassed by recommendation #1 in a single report in every case is impractical. Instead, if not in the report, the documentation described herein must be maintained in a case record. Generic documentation such as standard operating procedures and definitions must either be a part of the case record or be easily accessible (e.g., posted on a Web site, available on request).*
  3. *Reports should clearly state: the purpose of the examination or testing; the method and materials used; a description or summary of the data or results; any conclusions derived from those data or results; any discordant results or conclusions; the estimated uncertainty and variability; and possible sources of error and limitations in the method, data, and conclusions.*
  4. *Every report should state that the report does not contain all of the documentation associated with the work performed. In order to understand and evaluate all the work performed, and*

independently analyze and interpret the data and draw conclusions, a review of the case record is required.

5. *The case record should be organized and made available in a manner consistent with the National Commission on Forensic Science discovery recommendations.*

9.2 Process Analysis

A forensic biology laboratory should develop its own procedures based upon their own validation studies and results reported in the literature. Practical experience with instrumentation and results from performing casework are also important factors in developing processes for analysis. Standard operating procedures should be developed at the laboratory through empirical study. Validation studies are performed of each method to determine the lab specific applications and parameters. Each laboratory must follow proper documented procedures for analyzing evidence in order to minimize loss and contamination, and to obtain accurate results based on best practices. These standard operating procedures must be reviewed annually by the technical leader in the lab. The procedures are to include information such as "reagents, sample preparation, extraction methods (to include differential extraction of nuclear DNA samples with adequate amount of sperm), equipment, and controls which are standard for DNA analysis and data interpretation."[1] Reagents used must be labeled and documented correctly. These procedures are subject to annual audits by accrediting bodies and inspection by the FBI as a condition for CODIS participation.

9.2.1 Considerations of Analytical Limitations

Just like any forensic discipline, biology can be extremely challenging and it comes with limitations. DNA profiles taken from crime scenes are often dirty and degraded. Complex mixtures are more and more prevalent and these take significant time to interpret. Given the complexities facing the discipline, it is even more important for analysts to understand the limits of their work. These are largely established by validation, but there should be a point when a result is too partial or too complex to make an interpretation. The laboratory should define which data is interpretable, and of that interpretable data, which data should be deemed inconclusive.

9.2.2 Required Testing Controls

To ensure the results produced from analysis are accurate, controls must be utilized during the forensic biology testing process. In DNA analysis, each sample should be run with a positive control and a negative control,

DNA Testing Process Part 1

or reagent blank. A positive control contains all of the reagents for a reaction, and a known DNA profile to demonstrate that all the equipment and reagents are working properly. If the result produced for this standard sample is incorrect, incomplete, or missing, it indicates a problem. A negative control is a tube containing reagents only, but no DNA. The term "reagent blank" is sometimes used synonymously with "negative control." These controls should be run next to the samples at all stages of the analysis, using the same consumables employed to process the samples. Negative controls or reagent blanks should be run at the most stringent conditions in which corresponding samples are run. For example, if an evidence sample is concentrated down to 10ul and all 10ul are inputted into the amplification system, the negative control should be concentrated and amplified at the same stringency. If any DNA results come back in the negative control or reagent blank, this indicates a contamination event has occurred.

9.3 Reporting

9.3.1 Requirements

There are a number of standards from the FBI QAS, ANAB, and ISO that address what is required to be included in a forensic DNA report. Reporting requirements are covered in Section 11 of the FBI QAS:

Standard 11.1: The laboratory shall have and follow written procedures for taking and maintaining casework notes to support the conclusions drawn in laboratory reports. The laboratory shall maintain all analytical documentation generated by analysts related to case analyses. The laboratory shall retain, in hard or electronic format, sufficient documentation for each technical analysis to support the report conclusions such that another qualified individual could evaluate and interpret the data.

Standard 11.2 states that casework reports must include:

- Case identifier
- Description of evidence examined
- A description of the technology
- Locus or amplification system
- Results and/or conclusions
- A quantitative or qualitative interpretative statement
- Date issued
- Disposition of evidence
- A signature and title, or equivalent identification, of the person accepting responsibility for the content of the report.

192   Guide to the ABC Biology Exam

These reports must all be confidential and only be released under certain pre-approved and documented circumstances.

Laboratory reports must also be reviewed and this review must be documented "to ensure conclusions and supporting data are reasonable and within the constraints of scientific knowledge."[2] The review must include the following as listed in Standard 12.2 of the QAS:

12.2.1 A review of all case notes, all worksheets, and the electronic data (or printed electropherograms or images) supporting the conclusions.

12.2.2 A review of all DNA types to verify that they are supported by the raw or analyzed data (electropherograms or images).

12.2.3 A review of all profiles to verify correct inclusions and exclusions (if applicable) as well as a review of any inconclusive result for compliance with laboratory guidelines.

12.2.4 A review of all controls, internal lane standards, and allelic ladders to verify that the expected results were obtained.

12.2.5 A review of statistical analysis, if applicable.

12.2.6 A review of the final report's content to verify that the results/conclusions are supported by the data. The report shall address each tested item or its probative fraction.

12.2.7 Verification that all profiles entered into CODIS are eligible, have the correct DNA types and correct specimen category.

In addition, documentation of this review should be contained in the case record.

9.3.2 Quantitative/Qualitative Conclusions

When reporting the conclusions of DNA testing, statistical calculations are used to assign a weight to the association between an individual and an evidence profile. There are three main ways this is done: the random match probability (RMP), the combined probability of inclusion (CPI), and the likelihood ratio (LR). The RMP is used for single source profiles or profiles where genotypes have largely been deduced. When the interpretation is based upon the assumption of a single contributor (or a single major or single minor contributor to a mixture), the RMP formulae are those described in NRC II Recommendations 4.1, 4.3, and 4.4. For heterozygote genotypes, the formula is 2pq. For homozygote genotypes, the formula is $p^2 + p (1 - p) \theta$, where typically $\theta = 0.01$ (for most U.S. groups) or 0.03 (for some isolated populations) as recommended by NRC II.

The CPI is used for mixed profiles where genotype combinations cannot be deduced. Both the RMP and CPI calculate the chance that a randomly selected, unrelated individual would match or be included in the evidence profile. The CPI is the product of the individual locus PIs: (CPI = PI1 × PI2 × …). The probability of exclusion (PE) for a locus has been commonly presented two ways:

DNA Testing Process Part 1

PE = 1 − PI, or PE = q^2 + 2pq, where p is the sum of allele frequencies and q represents all other alleles (1 − p). Population substructure corrections can also be applied using: PE = 1 − [p^2 − p(1 − p)θ], where p is the sum of allele frequencies observed at that locus. The CPE has been commonly presented two ways: CPE = 1 − CPI, and CPE = 1 − [(1 − PE1) × (1 − PE2) × ... × (1 − PEN)].

The LR is the probability of observing the evidence if the suspect contributed to the DNA divided by the probability of observing the evidence if a randomly selected, unrelated person was the contributor. The equation for the likelihood ratio is $LR = \dfrac{Pr_1}{Pr_2}$.

The three methods described above are all considered "binary" approaches because they are applied based on the presence or absence of allelic information. These methods should only be applied to DNA profiles where there is no likelihood of allelic dropout or missing information. A fourth method for applying a statistical weight, probabilistic genotyping (Chapter 8), utilizes the likelihood ratio in its analysis, but also includes calculations for information that might be missing.

After an evidence DNA profile has been generated and assessed for interpretability, a suspect profile can be compared. The possible results can be:

Match – The peaks in the evidence profile and the peaks in the known profile have the same genotypes with no unexplainable differences. Statistical evaluation of the significance of the match should be reported. The strength of the association between the suspect's profile and the evidence profile must be expressed by use of statistical calculations. The calculations are designed to provide a lay judge or juror with an idea of how common or rare a DNA profile would be in a population.

Exclusion – The peaks in the evidence profile and the peaks in the known profile demonstrate differences that can only be explained by the two samples originating from different sources. No statistical evaluation is performed when an exclusion is reported.

Inconclusive – The data does not support a conclusion either way because it is not of sufficient quality. In this event, because there is no positive association between the evidence and a person of interest, no statistical evaluation is necessary.

9.4 Artificial Intelligence

9.4.1 Second Read Software and Automation

With the large quantity of biological samples that need to be processed, automated technologies are being used to keep up with the casework load. Expert systems and probabilistic genotyping software are discussed in more detail in Chapter 8.

9.5 Visualization Tools/Techniques

9.5.1 Microscopy

Microscopes are instruments that magnify an object and resolve the minute details. Early forensic scientists relied solely on microscopic examination of physical evidence, and even with the advanced technology that exists today, the microscope is still a useful asset in forensic analysis. In a microscope, or in any case when something is being magnified by a lens of glass, there is a real image and a virtual image. The real image is what is seen directly looking at the object, and the virtual image is what is seen only through a lens.

The microscope has many components vital to its function. The objective lens is the lower lens on a microscope, which is placed directly over the sample. Typically a microscope will have multiple objective lenses at 4×, 10×, 20×, and 45× magnification. The eyepiece lens, also known as the upper lens or ocular lens, is where the viewer looks into the microscope. The eyepiece lens typically has a magnification factor of 10×. The sample is magnified through each lens, objective and eyepiece, for a combined magnification of 40×, 100×, 200×, or 450× depending on which objective lens is being used. The base is where the instrument rests. The arm is attached to the base standing upright to support the microscope. The stage is the plate where the samples are placed to be studied, typically on glass slides that are held by stage clips. The body tube is a hollow tube connecting the objective and eyepiece lenses where the light passes through from the sample to the eye. The coarse adjustment is a knob that focuses the lens by raising and lowering the stage or the body tube. The fine adjustment knob is similar to the coarse adjustment but it moves the sample at a much smaller magnitude. An illuminator supplies artificial light from beneath the stage. A condenser collects all of the rays of light from the illuminator and concentrates them on the sample with even illumination.

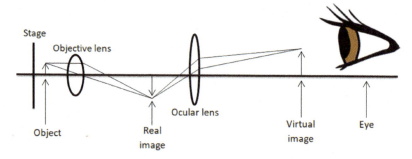

Figure 9.3 An illustration of the real and virtual images on a microscope.

Multiple types of microscopes are utilized for different forensic purposes. The compound microscope is the best known form, which magnifies a single sample through the devices defined above. This kind of microscope is used for identification of sperm cells in a semen sample, hair and fiber analysis, and analysis of paint chips.

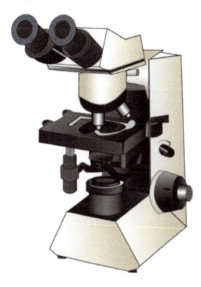

Figure 9.4 Compound microscope.

A comparison microscope allows side-by-side comparison of two samples. Through a series of mirrors, this microscope will create two images side by side that can be seen through a single ocular lens. Comparison microscopes can be utilized for comparison of bullets, cartridges, hairs, or fibers.

Figure 9.5 The two separate stages of a comparison microscope.

The stereoscopic microscope, or stereoZoom, provides magnification on a lesser scale than a compound microscope, ranging from 10× to 125× magnification. Stereoscopic microscopes create a three-dimensional image of the object through the alignment of two compound microscopes into one ocular lens. Stereoscopic microscopes are utilized in forensics for locating stains or trace evidence on garments, weapons, or tools. They are also useful for paint, soil, gunpowder residues, and marijuana analysis.

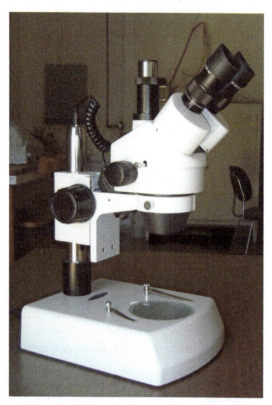

Figure 9.6 Stereoscopic microscope.

A polarizing microscope operates on the knowledge that light vibrates. Light passing through special crystal structures emerges vibrating on one plane, or as plane-polarized light. A polarizer is the material that emits the plane-polarized light, and an analyzer is a detector for this type of light. In a polarizing microscope, a stereoscopic microscope is outfitted with a polarizer and an analyzer so the viewer can detect any polarized light. Polarized microscopes are useful for examining minerals in soil and crystals.

A phase contrast microscope utilizes the differences in the refractive indices of different organelles to distinguish them. By adding filters above and below the condenser on a compound microscope, the light waves entering and leaving the sample are adjusted. Refractive index is measured as the

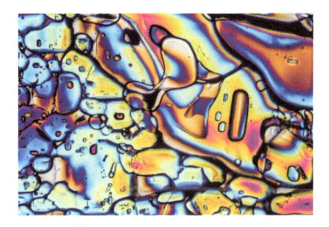

Figure 9.7 Ammonium sulfate crystals illuminated with polarized light under a microscope.

speed of light in a vacuum divided by the speed of light in a medium. Different organelles have different refractive indices, and thus will bend, refract, or reflect light in different ways, causing light waves to be 'out of phase' with one another. This resulting image appears like that of a negative film strip, where objects that are normally light appear dark, and vice versa. Phase contrast can be used to help identify sperm cells on a microsocope slide.

Automation of microscopic identification is slowly beginning with an instrument known as the microspectrophotometer. This device links the microscope to a computerized spectrophotometer. This method is useful in

Figure 9.8 A microspectrophotometer.

many ways, but particularly with the provision of information to characterize trace amounts of evidence, distinguish counterfeit forms of currency, and examine fibers and paints.

The scanning electron microscope (SEM) is unique in its formation of the image of the specimen. Rather than use the light coming off of the sample to view it, the SEM aims a beam of electrons onto the sample and examines the electron emissions. The electrons are collected and displayed on a monitor. SEM magnifies the sample up to 100,000× with high resolution and great depth of field. SEM is useful for a range of applications, including identifying counterfeit currency and marijuana, gunpowder residues, and investigating spores and pollen samples.

Figure 9.9 Scanning electron microscope.

9.5.2 Electrophoresis

Electrophoresis is a method for the separation of molecules via migration in a support medium using electrical potential. Electrophoresis is useful in forensic biology for separating and identifying proteins and DNA, and analyzing complex biochemical mixtures.

9.5.2.1 Gel Electrophoresis

Today, the most commonly separated molecule in forensic biology is DNA. DNA has an overall negative charge from the phosphate groups present in the DNA backbone. Therefore, DNA will migrate in an electrical field from the negative electrode (cathode) to the positive electrode (anode).

Electrical potential is a measure of the energy required to move a molecule that is charged in an electric field. Electrophoretic mobility is determined by charge-to-mass ratio, shape of the macromolecule, and size of the macromolecule. With DNA, since every phosphate group carries a negative

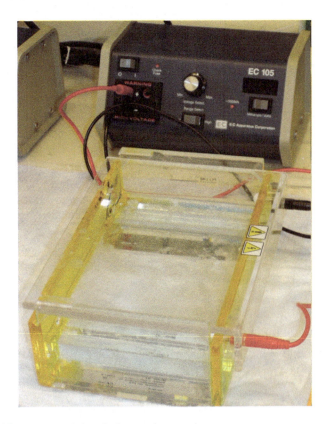

Figure 9.10 Agarose slab gel electrophoresis apparatus.

charge, the charge-to-mass ratio is the same for all DNA fragments regardless of length. Also, DNA shape is typically linear during electrophoretic analysis, either double-stranded or single-stranded. DNA separation through electrophoresis is due to the varying sizes of each fragment.

The material for the supporting matrix for electrophoresis is commonly an agarose or polyacrylamide gel. Agarose is preferred for separation of larger fragments (50–20,000 base pairs), and polyacrylamide for smaller (5–500 base pairs). Agarose gel is prepared as a slab with wells on one end for the DNA to be inserted; polyacrylamide gel can be prepared in slab form or in a capillary electrophoresis apparatus. These gels are composed of a matrix with tiny pores through which the DNA travels. Naturally, the smaller DNA fragments can more easily maneuver through these small pores, so the smaller a fragment, the quicker it moves through the gel. In other words, electrophoretic mobility increases as DNA size decreases.

In horizontal slab electrophoresis, the gel matrix is heated to liquid form, poured into a cast, and allowed to cool to a solid gel. The gel contains small wells on one end for the DNA samples. It is submerged into a liquid buffer and the DNA is loaded into the wells at the cathode. The DNA samples are often mixed with a dye that will not interfere with the DNA migration so as to facilitate the loading of the DNA into the wells and to monitor the progression of electrophoresis, since the dye travels faster than any DNA fragment. Multiple samples can be run simultaneously in a slab gel. Positive controls of a mixture of known DNA fragment lengths should be run alongside samples to determine the length of the sample fragments based on the positive control. After the electrophoresis is complete, the samples are made visible through a fluorescent reagent that makes DNA visible under UV light, or by staining the gel with a chemical such as ethidium bromide or Coomassie Blue.

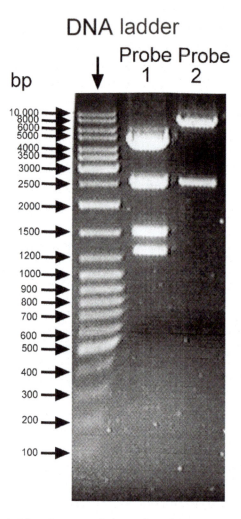

Figure 9.11 DNA ladder shown on left used to determine size of fragments in the samples.

DNA Testing Process Part 1 201

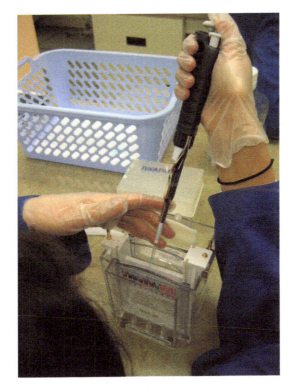

Figure 9.12 Polyacrylamide gel electrophoresis apparatus.

Figure 9.13 Coomassie Blue stain of DNA in gel electrophoresis.

Polyacrylamide slabs are run in a vertical formation rather than horizontal. Polyacrylamide gel electrophoresis is ideal for STR fragment separation and mtDNA sequencing products. Polyacrylamide gel can separate fragments with one base pair change, so this medium is ideal for sequencing DNA.

9.5.2.2 Capillary Electrophoresis

Capillary electrophoresis can be used for the separation of DNA, RNA, polysaccharides, and proteins, and is useful in forensics for STR analysis and mtDNA sequencing. The sample is injected into a thin capillary tube filled with a matrix of linear polydimethylacrylamide using an electrokinetic injection. This capillary is connected to buffer reservoirs that are themselves connected to electrodes. DNA fragments travel again to the positive electrode and are separated according to their sizes with single nucleotide resolution. At the end of the capillary tube is a detection system that excites the fluorescently labeled DNA fragments with a laser and then records the signals emitted. Capillary electrophoresis is a rapid process but each capillary can only process one sample at a time.

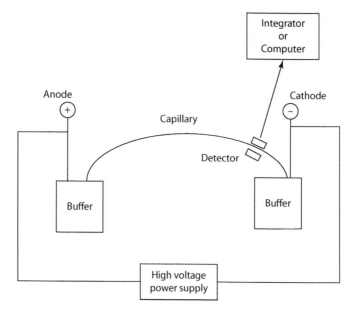

Figure 9.14 Capillary electrophoresis.

Figure 9.15 This is an example of a Powerplex Fusion Ladder run through a capillary electrophoresis instrument.

DNA Testing Process Part 1   203

Migration of the sample through the capillary is due to two types of mobility: electrophoretic mobility and electroosmotic mobility. Electrophoretic mobility is the solute's response to the applied electric field. Cations move toward the negatively charged cathode, anions move toward the positively charged anode, and neutral species, which do not respond to the electric field, remain stationary. The other contribution to a solute's migration is electroosmotic flow, which occurs when the buffer solution moves through the capillary in response to the applied electric field. Under normal conditions, the buffer solution moves toward the cathode, sweeping most solutes, even anions, toward the negatively charged cathode.

9.5.3 Fluorescence

Fluorescence is a method used for the identification of different substances. In forensic biology, we rely on fluorescence to detect DNA during real-time PCR–based quantitation methods, as well as DNA fragment detection in capillary electrophoresis. Moreover, fluorescence is detected with the use of an alternative light source to help identify stains, and it can be used in some presumptive tests for body fluids. Fluorescence occurs when an object exposed to a particular wavelength is able to reemit it in a different, typically visible light wavelength. Fluorescence detection methods have greatly aided the sensitivity and ease of measuring PCR-amplified STR alleles. In real-time PCR quantitation, the system is measuring the cycle-to-cycle change in fluorescence. In capillary electrophoresis, pieces of DNA are tagged with fluorescent dyes that are detected by a CCD camera located just outside a window in the capillary. We visualize those pieces of DNA as peaks that are measured in relative fluorescence units (RFU). The primary instrument platform used

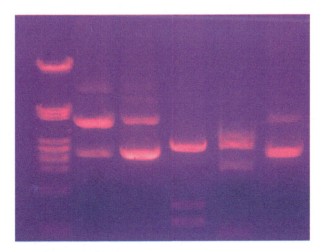

Figure 9.16 Fluorescent detection of DNA separated in gel electrophoresis.

Guide to the ABC Biology Exam

in the United States for fluorescence detection of STR alleles is currently the ABI 3500 Genetic Analyzer.

Notes

1. FBI QAS 9.1.1.
2. FBI QAS 12.1.

Bibliography

Butler, J. (2005). *Forensic DNA typing* (2nd ed.). Elsevier.

Li, R. C. (2015). *Forensic biology* (2nd ed.). Boca Raton: CRC Press (Ch 7, 8, 25).

Nat'l Comm'n on Forensic Sci., Meeting #13: April 10–11, 2014 2 (2017). https://www.justice.gov/ncfs/page/file/976566/download

Nat'l Comm'n on Forensic Sci., Views of the Commission: Report and Case Record Contents (2017). https://www.justice.gov/ncfs/page/file/952696/download

Saferstein, R. (2018). *Criminalistics: an introduction to forensic science* (12th ed.). Boston: Pearson (Ch 7,15).

SWGDAM. Interpretation Guidelines for Autosomal STR Typing by Forensic DNA Testing Laboratories – APPROVED 01/12/2017.

DNA Testing Process Part 2

10

10.1 Isolation and Purification of Nucleic Acids

DNA extraction is a process in which cells are broken open to release the DNA from the nucleus and, at times, the DNA is purified. DNA extraction is necessary to free the DNA from the nucleus to allow it to be processed further. Because there are many other substances in the cell besides DNA, such as proteins and organelles designed to protect the DNA and carry out other cellular functions, sometimes these are first cleaned away before the DNA is examined. Current methods of DNA extraction can vary depending on the type of biological evidence and the test that will be utilized following the extraction. The goal of all DNA extraction methods is to recover the highest yield of pure, high-quality DNA that will produce a complete DNA profile.

The general method for DNA extraction is similar for all protocols. Primarily, the tissue and cells in which the DNA exists must be disrupted through use of enzymatic digestions, such as proteinase K, by boiling, or using alkali treatment. Bone and teeth samples are frozen with liquid nitrogen and ground to a powder and then decalcified using a chelating agent.

Either in the process of tissue disruption or right afterward, membranes of the cells, nuclei, and mitochondria are all lysed to release both forms of DNA, nuclear and mitochondrial. Lysis, or breakage, of cells is performed with salts and chaotropic agents including detergents such as sodium dodecyl sulfate (SDS) and sarkosyl.

The process of lysing membranes also denatures proteins and dissociates the histone proteins from DNA. TE buffer, a combination of Tris-HCl and EDTA (ethyldiaminetetraacetic acid), is usually employed as a buffer medium in which the lysis takes place to maintain a pH level in which endogenous deoxyribonucleases (DNases) remain inactive. DNases are enzymes that catalyze the breaking of phosphodiester bonds in DNA, so it is crucial that they remain inactivated. Chelating agents including EDTA or Chelex can be used to inactivate DNases as they chelate, or bind up, the cations that work together with DNases, since without the cations the DNases don't perform the lysis of phosphodiester bonds.

After membrane lysis comes the removal of proteins and liquids that interfere with DNA extraction. This can be done through extraction with organic solvents or reverse binding of DNA to a solid such as silica, and removing interfering species through washing. At this point, the DNA has

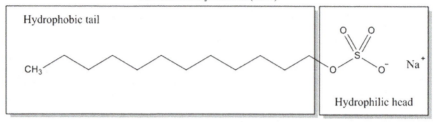

Figure 10.1 SDS is a chaotropic agent used to lyse cells for DNA extraction.

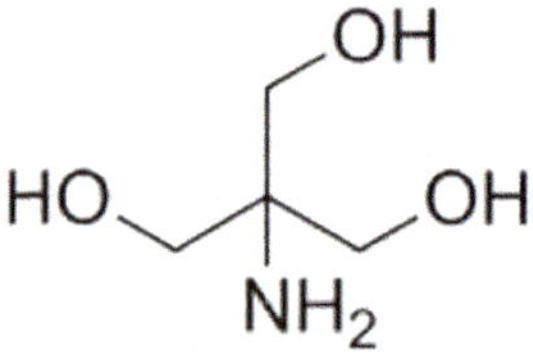

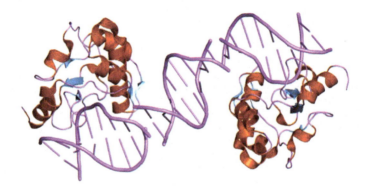

Figure 10.2 TE buffer helps maintain a stable pH between 7.5 and 9, which helps to stabilize the DNA over a longer period of time and prevent degradation.

Figure 10.3 DNases breaking down DNA.

been released from the cell and purified. From here, the DNA is typically stored in TE buffer at pH 8. Freezing the sample is recommended for long-term samples, but freezing and thawing cycles should be avoided as the rapid temperature differences might cause DNA breakage. There are four major extraction methods currently used in forensic DNA analysis: organic extraction, differential extraction, Chelex extraction, and silica-based extraction.

10.1.1 Organic Extraction

Solvent-based extraction, also known as organic extraction, is unique in its utilization of phenol–chloroform. Cell lysis and protein digestion occurs with proteinase K. Protein removal is carried out by mixing the DNA sample

DNA Testing Process Part 2

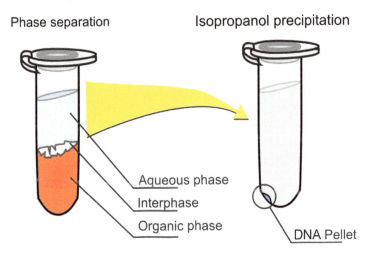

Figure 10.4 Separation of DNA and proteins using solvent-based extraction.

with phenol : chloroform : isoamyl alcohol in a ratio of 25 : 24 : 1. This mixture forms an organic phase in the bottom of the tube, which is easily separated from the DNA-containing aqueous supernatant.

DNA is concentrated using either ethanol precipitation or ultrafiltration. This method yields large, double-stranded DNA which is commonly analyzed with restriction fragment length polymorphism (RFLP) or polymerase chain reaction (PCR). There is no limit to the amount of DNA that can be extracted using this protocol. Unfortunately, organic extraction is time consuming and requires sample transfer among tubes, which could result in contamination or lab error. Additionally, phenol : chloroform is a hazardous chemical and its storage and disposal in the laboratory are costly.

10.1.2 Differential Extraction

Differential extraction is used to extract DNA from evidence found in sexual assault cases, including vaginal swabs and body fluid stains. Such samples typically contain mixtures of the DNA profiles of the victim and perpetrator, which must be separated for identification. Differential extraction is useful when the male DNA exists in the form of a sperm cell, and the male DNA is in low concentrations compared to the female DNA. The nonsperm and spermatozoa cells are lysed in separate steps because of the differences in the cell membranes of sperm cells versus those of nonsperm cells. The sperm cell membrane is much tougher than the cell membranes of other epithelial and somatic cells.

The first step involves lysing the weaker cell membranes of the nonsperm cells using proteinase K, collecting the supernatant containing nonsperm DNA, and placing that supernatant into a separate tube. This sample becomes the nonsperm fraction and will produce a DNA profile of predominantly nonsperm cells.

Next, the sperm cells are lysed through addition of dithiothreitol (DTT) and then extracted. This fraction becomes the sperm cell fraction because it is expected to produce a DNA profile predominantly from the sperm cell donor. This is not a perfect process since the two cell types don't often separate completely, and some cross over still occurs. However, this process is extremely beneficial in cases where sperm cells are present in very low amounts compared to epithelial cells.

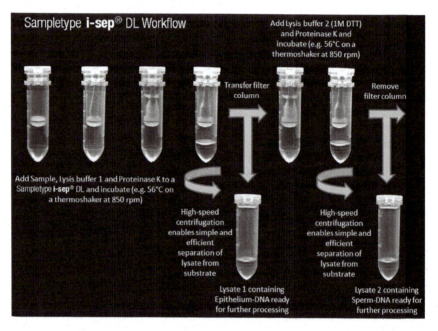

Figure 10.5 Differential extraction.

10.1.3 Chelex Extraction

Chelex extraction, also called boiling extraction, has been around since the early 1990s. First, contaminants are washed out of the sample, and the purified cells are incubated in a 5% Chelex solution at 56°C, a temperature chosen for its inactivation of DNases. The sample is incubated for 20 minutes to separate cells grouped together and to soften membranes. Cells are lysed when they are heated to boiling at 100°C. The Chelex solution inhibits DNase activity by binding to metal ions that are cofactors to DNases, such as magnesium. Centrifugation then separates the Chelex and cell waste to the bottom of the tube. The supernatant containing the DNA is now pure and ready for analysis. Chelex extractions use only one tube, reducing the risk of contamination or lab error; however, the boiling step denatures the DNA into fragmented single strands that are unsuitable for RFLP and can only be used for PCR.

Figure 10.6 Test tubes incubating in a hot water bath as they would during the boiling phase of Chelex extraction.

10.1.4 Silica-Based Extraction

The silica-based extraction process is a solid phase extraction technique involving the adsorption of DNA molecules to a silica surface on the pretense that DNA can be desorbed when surrounded with chaotropic salts. The cells are first lysed with proteinase K and then the released DNA adsorbs onto the solid silica. Other proteins and contaminants do not adsorb to the silica and are washed away at this time using an ethanol-based solution. The adsorbed DNA is then eluted from the silica. Silica-based extraction can be performed for high-throughput methods to process many samples at once, and it can be automated. However, the resulting DNA is in a single stranded form, and the amount of DNA recovered is limited to the binding capacity of the silica beads.

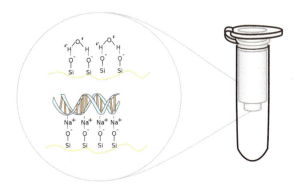

Figure 10.7 DNA adsorbing to silica in silica-based DNA extraction.

10.2 Quantification

Accurate quantification of DNA samples following extraction is vital for the success of downstream analysis. Quantification methods must be as specific for human DNA as possible while ignoring any other nonhuman DNA that may be present, such as microbial DNA. Three methods of DNA quantitation include slot blot, intercalating dye, and quantitative PCR (qPCR). The latter is the most commonly used method in crime labs today.

10.2.1 Slot Blot Assay

Slot blot assays can detect human and higher primate DNA and they rely on a few basic steps. First, the DNA is mixed with an alkaline solution to denature the DNA. This is necessary for cross linkage of the single-stranded DNA onto a nitrocellulose membrane. Next, a slot blot device is used to spot the DNA onto the nitrocellulose membrane, then to immobilize the DNA onto a nylon membrane. A labeled probe complementary to a primate-specific DNA sequence at the D17Z1 locus hybridizes to the target sequence, typically located in the centromere of chromosomes. The target sequence can then be detected in three different ways. The probe can be labeled with radioisotopes so the complementary strand to the probe can be seen through X-ray film. Since radioisotope detection can be hazardous, this method is not often used. A second method is to label the probe instead with alkaline phosphate, which is visible through chemiluminescent detection. The third method is to use a biotinylated probe that can be detected through either colorimetric or chemiluminescent techniques.

Colorimetric DNA detection involves binding the biotin probe to a streptavidin complex bound to horseradish peroxidase. This catalyzes the oxidation of tetramethylbenzidine (TMB) to form a blue precipitate.

Chemiluminescent detection is more sensitive than colorimetric. It also uses horseradish peroxidase to oxidize substrates such as luminol, emitting photons that can be detected using X-ray film. The intensity of the outputted signal is proportional to the DNA concentration, which is calculated through comparison to a set of known standards. Slot blot assays handle quantitation of DNA ranging from 150pg to 10ng. The major limitation to slot blot assays is that the results are read manually, and therefore some subjectivity exists in the determination of the quantity of DNA present. Additionally, they consume a large amount of sample. These "yield gels" depict the quantity of DNA based on the output of whichever method is being used.

10.2.2 Intercalating Dye Assay

Intercalating dye assays determine DNA quantities fluorescently, detecting up to 250pg of DNA. These dyes are extremely small planar molecules that insert

DNA Testing Process Part 2

themselves between the base pairs of DNA without disrupting the hydrogen bonds. The DNA sample is added into the fluorescent intercalating dye and the wavelength is measured in a spectrofluorometer. Since the fluorescence is proportional to the amount of DNA, the wavelength outputted can be compared to a standard curve to determine the quantity of DNA present. Intercalating dyes bind to all DNA molecules, not just human DNA, so they are best used for quantitation of known reference standards. This assay is also amenable to automation and can process a high quantity of samples rapidly. Intercalating dyes such as SYBR green can also be utilized in qPCR.

10.2.3 Quantitative PCR

The qPCR method is the most sensitive of all the DNA quantitation methods, detecting human and nonhuman primate DNA with high rates of success. The details of qPCR are explained in Chapter 8, but a few concepts will be expanded upon here. In the real time-PCR method, probes are labeled with two fluorescent dyes that emit light at different wavelengths.

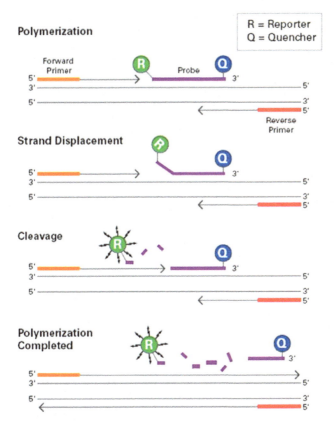

Figure 10.8 TaqMan probe, qPCR method.

The probe sequence is designed to attach specifically in the target region between the two primers. A reporter dye is attached at the 5′-end of the probe sequence, while a quencher dye is synthesized on the 3′-end. When the probe is intact and the reporter dye is in close proximity to the quencher dye, the signal of the reporter dye is suppressed. As the new strand is built, the DNA polymerase will begin to knock off any probes that have annealed to the target sequence. When the reporter dye molecule is hydrolyzed during DNA extension, it will begin to fluoresce. That fluorescence is measured and compared to a standard curve in order to estimate the quantity of DNA in the sample. The Quantifiler kits from Applied Biosystems as well as the PowerQuant® system from Promega utilize this method of qPCR.

This process has high accuracy when handling small DNA quantities and it is amenable to automation, which eliminates more opportunities for human error. Moreover, real-time qPCR allows for the detection of more than one type of DNA target sequence during the same reaction, such as nuclear DNA and Y-chromosomal DNA or mitochondrial DNA.

10.3 Polymerase Chain Reaction

Quantification of DNA is vital for successful PCR since only a small range of DNA can be used for accurate amplification. Too much DNA inserted will cause interference from artifacts and complicate data analysis. Too little DNA can result in a partial profile or no profile at all. The details of PCR are explained in Chapter 8.

10.4 DNA Typing Technology

10.4.1 Fragment Analysis/Short Tandem Repeat Analysis

The DNA testing process begins with DNA extraction and quantification through one of the methods described above. From here, the DNA is amplified using PCR and separated via electrophoresis. The PCR products are detected and the sizes of short tandem repeat (STR) fragments are determined. Finally, an electropherogram is made from the data using a software program, and the interpretation process begins.

10.4.1.1 Theory

More than 99% of human DNA is not unique from person to person. We have genes for hair color and eye color, and when we look around we can tell those traits are not unique to us. But the combination of STRs in our DNA is unique. The entire forensic DNA process is designed to examine these short segments of DNA that vary in length. Those lengths are translated by

DNA Testing Process Part 2

Table 10.1 Example of a Single-source Profile Match

Locus	Single-source DNA Profile from Evidence	Single Source Suspect Profile
D8S1179	12, 13	12, 13
D21S11	28, 29	28, 29
D7S820	12, 12	12, 12
CSF1PO	9, 11	9, 11
D3S1358	8, 9	8, 9
TH01	7, 9.3	7, 9.3
D13S317	12, 13	12, 13

computer software and allelic ladders into a series of numbers: a person's own unique numerical code. The evidence will also have its own "code" or profile. After the data has been examined, DNA peaks are differentiated from artifacts, genotypes are determined, and we look to see if the numbers match.

10.4.1.2 Application/Processes

For STR profiling to be successful, there should be no circumstances where two random DNA profiles match on all STR regions studied. Therefore, it is most useful if the STRs that are being compared are complex and highly variable among the population. The STR regions should not be linked, but ideally located far apart on a chromosome or even on different chromosomes. In STR analysis, the STR loci chosen are labeled with fluorescent primers and amplified. The amplified STRs are separated using electrophoresis, and the fluorescent dyes are detected. The signals outputted will correspond to each DNA fragment and are recognized using a computer software program and organized into an electropherogram, a diagram showing a profile of peaks that each correspond to a DNA fragment. The peaks are located in order of electrophoretic mobility—in other words, in size order from smallest to largest. A standard is used to compare the DNA fragments to determine the size of each. The area of the peak demonstrates the fluorescent signal intensity in relative fluorescence units (RFU). The data in an electropherogram is plotted as fluorescent signal intensity (in RFU) versus the size of the DNA fragments.

From the electropherogram, the analyst must determine what peaks are true DNA peaks, and which, if any, are artifacts. This means that stutter peaks, dye blobs, and spikes are removed from the allelic data set. When only allelic information remains, meaning the analyst believes, using validated metrics, that all remaining peaks are true DNA, other qualitative assessments of the profile occur.

10.4.1.3 Interpretation/Results

The electropherogram provides the data from the STRs that can be converted into a genotype. The genotype represents the combination of alleles at any given locus. This is fairly easy when the DNA profile is the result of a single source sample. It involves removing artifacts and simply recording the allelic information present in the profile. This data is easily recorded and stored in databases such as CODIS. Allelic ladders are the key to producing accurate allele calls on an electropherogram. An allelic ladder is a grouping of synthetic fragments that correspond to alleles that are common in humans for specific STR loci. Think of the allelic ladder like a genetic measuring tape or ruler that measures the length of each DNA fragment. Comparing the allelic ladder and the DNA fragment determines the size of STR alleles. Rare alleles that do not match an allelic ladder are known as off-ladder alleles, or microvariants.

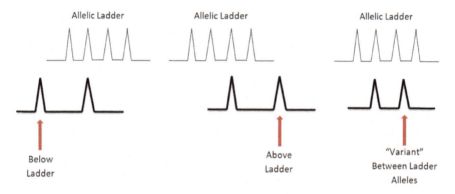

Figure 10.9 Three types of "off-ladder" alleles.

Interpreting the STR profile results is a complex process that has been outlined by the Scientific Working Group on DNA Analysis Methods (SWGDAM) and the DNA Commission of the International Society of Forensic Genetics (ISFG). It involves differentiating true DNA peaks from instrument noise using an analytical threshold, identifying non-allelic peaks, determining the overall quality of the profile (to include using a stochastic threshold for suitability for binary statistical calculations), and assessing the number of potential contributors. If the profile is suitable for comparison, the analyst moves on to determine if similarities exists between the evidence DNA profile and the DNA profiles from persons of interest. As discussed in previous chapters, results are categorized as inclusion, exclusion, or inconclusive. Inclusion indicates similarity between the STR peaks of the person of interest with the crime scene evidence. Exclusion indicates that the genotypes are different and the samples are from two separate sources. Inconclusive results indicate that the data does not support inclusion or exclusion due to insufficient information. For specific instruction and examples on interpretation,

DNA Testing Process Part 2

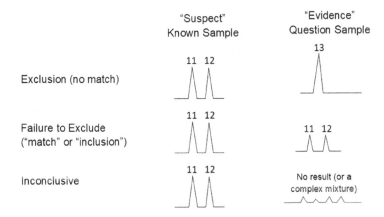

Figure 10.10 Exclusion, inclusion, and inconclusive results for DNA analysis.

reference the most recent version of the SWGDAM Interpretation Guidelines for Autosomal STR Typing by Forensic DNA Testing Laboratories and also *Forensic DNA Typing: Interpretation* by John M. Butler.

Bibliography

Li, R. C. (2015). *Forensic biology* (2nd ed.). Boca Raton: CRC Press (Ch 3, 5, 6, 20).

Index

β-amylases, 93
4-methylumbelliferone phosphate (MUP), 91
17-ketosteroids, 96

ABAcard HemaTrace®, 84, 163
AB blood, 81
ABC, *see* American Board of Criminalistics
ABFDE, *see* American Board of Forensic Document Examiners
ABFT, *see* American Board of Forensic Toxicology
A blood, 80
ABO blood typing, 6
Accreditation, 39–40
AccuTrans, 36
Acid phosphtase (AP), 89
Adenine, 107, 109
Adenosine triphosphate (ATP), 88, 133
Ad testificandum, 65
Affidavits, 65
AFTE, *see* Association of Firearm and Tool Mark Examiners
Agarose gel, 199
Agglutination, 81, 163
Alcohol, 20, 24
Alcohol, Tobacco, and Firearms labs, 19
Alkanes, 67
Alkenes, 67
Alkynes, 67
Alleles, 123, 124, 127, 128, 141, 148, 160
Allele-specific oligonucleotide (ASO), 136
Alternative Light Source (ALS), 73, 79
Ambien, 26
American Board of Criminalistics (ABC), 2, 42
American Board of Forensic Document Examiners (ABFDE), 43
American Board of Forensic Toxicology (ABFT), 43
American Society for Testing and Materials (ASTM), 50
American Society of Crime Laboratory Directors (ASCLD), 40

American Standard's Board (ASB), 50
Amphetamines, 25
Amplicon, 169
AmpliTaq Gold DNA polymerase, 172
ANAB, *see* ANSI National Accreditation Board
Anabolic steroids, 25
Aneuploidy, 137, 138
Animal cells, 71
Animal forensics, 150
ANSI National Accreditation Board (ANAB), 40
Anthropometry, 7
Antibodies, 81
Antigen, 162–164
    and immunological systems, 161–164
    polymorphisms, 163
Antihuman hemoglobin (Hb) antibody, 85, 86
AP, *see* Acid phosphtase
Applied Biosystems GlobalFiler kit, 176
Archiv für Kriminal Anthropologie und Kriminalistik, 3
Arch pattern of fingerprints, 31
Arenes, 67
Aromatic hydrocarbons, *see* Arenes
Arrhenius acid, 69
Arson evidence, 15
ASB, *see* American Standard's Board
ASCLD, *see* American Society of Crime Laboratory Directors
ASO, *see* Allele-specific oligonucleotide
Aspermia, 89
Association of Firearm and Tool Mark Examiners (AFTE), 43
ASTM, *see* American Society for Testing and Materials
Ativan, 26
ATP, *see* Adenosine triphosphate
Audits and auditors, 39
Audit trail, 40
Autosomal DNA, 158
Autosomal trisomy, 138
Azoospermia, 89

218 Index

Bacterium, 72
Ballistic evidence, 19
Barberio's test, 90
Barbiturates, 24, 25, 26
Bayes, Thomas, 75
Bayesian inference, 75
Bayes' theorem, 75
B blood, 80, 81
Bell, Suzanne, 75
Bertillon, Alphonse, 2
BFDE, *see* Board of Forensic Document
    Examiners
Biological evidence, 17, 79
Biological material, 10, 17, 20
Biological screening tests, 79–97
    blood, 79–87
        confirmatory tests, 84–87
        presumptive tests, 83–84
        species identification, 87
        typing, 80–82
    feces, 97
    saliva, 93–95
        confirmatory tests, 94–95
        presumptive tests, 93–94
    semen, 88–93
        components, 88–89
        confirmatory testing, 91–93
        presumptive tests, 89–91
    urine, 95–97
        confirmatory tests, 96–97
        presumptive tests, 96
        properties, 95–96
Black Pad case, 167
Blind proficiency test, 56
Blood, 79–88
    alcohol testing, 29, 73
    confirmatory tests, 84–87
    presumptive tests, 83–84
    protein analysis, 6–7, 73
    species identification, 87
    typing, 80–82
Bluestar, 84
B lymphocyte, 161–162
Board of Forensic Document Examiners
    (BFDE), 50
Body fluids, 6, 10, 17, 20, 29, 72, 73,
    79, 152
Boiling extraction, *see* Chelex extraction
Brady v. Maryland (1963), 62
Brentamine test, 90
Bromothymol blue, 96
Brønsted-Lowry acid, 69

Cambridge Reference Sequence (CRS), 136
Capillary electrophoresis, 72, 202–203
cDNA, *see* Complementary DNA
Cells
    and chromosomes, 103–106
    division, 108–109
    DNA content, 107–108
    morphology, 99–103
Cetus Corporation, 168
Chain of custody, 18, 39, 62
Chelex extraction, 208–209
Chemical hygiene, 56–57
Chemical mapping technique, 79
Chemical waste, 58
Chemiluminescent tests, 83, 84
Chloroform, 207
Choline, 90
Christmas tree stain, 91
Chromatin, 103
Chromosomes, 103–106, 114
Civil laws, 156
Class evidence, 19
Club drugs, 25
Cocaine, 25
Codeine, 22, 23, 25, 26
CODIS, *see* Combined DNA Index System
Collaborative Testing Services, 55
Colorimetric assays, 83, 89, 93, 94, 96
Combined DNA Index System (CODIS), 2,
    41, 160, 169, 181, 183, 190, 214
Combined probability of inclusion (CPI),
    192–193
Complementary DNA (cDNA), 178
Confidence interval, 53–54
Controlled substances analysis, 20–28
Corrective action, 55
Court testimony, 63–64
Covalent bond, 69, 70
CPI, *see* Combined probability of inclusion
"Crack," 25
Creatinine concentration, 96
Crick, Francis, 109
Crime scene
    documentation, 10–11, 12
    preservation, 8–14
Criminal investigation, 5, 6
Criminal justice system, 50
Criminal laws, 156
CRS, *see* Cambridge Reference Sequence
Cyanoacrylate fuming, 31
Cycloalkanes, 67
Cytogenetics, 137–138

Index

219

chromosomal abnormalities, 138
mitotic and meiotic, 137–138
Cytokinesis, 108, 109
Cytosine, 107, 109

DAB, *see* DNA Advisory Board
D antigen, *see* Rhesus factor
Databanking, 160
Date rape drugs, 25
Daubert v. Merrell Dow Pharmaceuticals
  (1993), 61, 62, 64, 65
Deductive reasoning, 76
"Denaturing protein electrophoresis," 165
De novo mutations, 119
Deoxyribonucleases (DNases), 205
Department of Justice, 188–189
Depressants, 24, 25
Developmental validation, 52–53, 184
Differential extraction, 207–208
Digital evidence, 36–37
Diploid cells, 123
Dipole-dipole interaction, 70
Dithiothreitol (DTT), 208
DMAC, *see* p-dimethylamino
  cinnamaldehyde
DNA, 13, 14, 58, 101, 103, 111, 114–116,
  167–170, 190; *see also specific*
  DNA entries
  amplification, 54, 89
  analysis, 20, 29, 40, 41, 72, 91, 150
  nucleotides, 171, 173
  polymerase, 171
  profiles, 18, 19, 41, 51, 72, 73–75, 97, 135,
    147–148, 160, 163, 182
  replication, 114–117
    enzymes involved in, 115–116
    forks and bubbles, 114–115
    organization, 114
    proofreading mechanisms, 116–117
  structure, 109–111
  testing, 155, 158, 176, 187
DNA Advisory Board (DAB), 40
DNA Control 007, 176
DNA evidence, 155–185
  CODIS, 184
  evolution, 160–183
    antigen and immunological systems,
      161–164
    DNA polymorphisms, 165–183
    protein and enzyme polymorphisms,
      164–166
  laboratory validations, 184

quality assurance standards (QAS)
  for DNA databasing laboratories, 183
  for forensic DNA laboratories, 183
  SWGDAM, 184
  types, 155–160
    criminal, 156–158
    databanking, 160
    kinship, 158–159
    missing-person/mass disaster, 158
    non-criminal, 158
DNA "fingerprinting," 167
DNA Identification Act (1994), 40, 160
DNA polymorphisms, 161, 167–183
  automation, 181–183
  genetic markers, 178–181
  PCR, 168–170
    cycle and its reaction components,
      171–174
    factors affecting, 174–177
  qPCR, 170–171
  real-time PCR, 170–171, 176
  RFLP, 167–168
DNases, *see* Deoxyribonucleases
DNA testing process, 6, 7, 187–204
  appropriate screening tests and
    examinations, 187–188
  artificial intelligence, 193–194
  case management, 187
  establishing case record, 188–190
  isolation and purification of nucleic
    acids, 205–209
    Chelex extraction, 208–209
    differential extraction, 207–208
    organic extraction, 206–207
    silica-based extraction, 209
  polymerase chain reaction, 212
  process analysis, 190–191
    analytical limitations, 190
    testing controls, 190–191
  quantification, 210–212
    intercalating dye assay, 210–211
    quantitative PCR, 211–212
    slot blot assay, 210
  quantitative/qualitative conclusions,
    192–193
  requirements, 191–192
  short tandem repeat (STR) fragments,
    212–215
    application/processes, 213
    interpretation/results, 213–215
    theory, 212–213
  visualization tools/techniques, 194–204

220   Index

electrophoresis, 198–203
fluorescence, 203–204
microscopy, 194–198
DNA transposons, 179, 180
Double-strand mutations, 120
Double-strand repair method, 120
Down syndrome, 138
Drug Enforcement Agency lab, 19–20
Drug evidence, 15
Drugs, 20
DTT, *see* Dithiothreitol
Duces tecum, 62, 65

Edelman test, 97
EDTA, *see* Ethyldiaminetetraacetic acid
Edwards syndrome, 138
Electrical potential, 198
Electromagnetic spectrum, 73
Electroosmotic mobility, 203
Electropherogram, 136, 182, 213, 214
Electrophoresis, 165, 198–203, 212
    capillary, 202–203
    gel, 198–202
Electrophoretic mobility, 199, 203
ELISA, *see* Enzyme Linked Immunosorbent
        Assays
Enantiomers, 71
Enzyme Linked Immunosorbent Assays
        (ELISA), 93, 94, 96
Enzyme polymorphisms, 161, 164–166
Eosin, 91
Erythrocytes, *see* Red blood cells
Ethics code, 50
Ethyl alcohol, 29
Ethyldiaminetetraacetic acid (EDTA), 205
Euchromatin, 106
Eukaryotic cells, 100, 103, 108
Expert systems, 182
Expert testimony, 61–62
Exterior packaging of evidence, 18

FBI, *see* Federal Bureau of Investigation
Feces, 97
Federal Bureau of Investigation (FBI), 19,
        40–41, 58, 63, 148, 160, 169, 181,
        183–184, 190, 191
Federal Controlled Substances Act
        (1970), 25
Federal Rule of Evidence 702, 61
Fingerprints, 4, 29–33
Firearm, 19
    analysis, 34–36, 43

and bullet markings, 6
Fire debris analysis, 33
Florence test, 90
Fluorescein, 84
Fluorescent technique, 79
Fluorescent test, 83, 84
Fluorometric assays, 89, 91
Footwear impression, 19
Forensic biology evidence, 156, 158, 160, 163
Forensic botany, 150–152
Forensic Chemistry, 77
Forensic Quality Services (FQS), 40
Forensic science, 2
    crime laboratory operations, 19–37
        biology, 20
        controlled substances analysis, 20–28
        digital evidence, 36–37
        fingerprints, 29–33
        firearms and toolmarks, 34–36
        fire debris, 33–34
        questioned documents, 33
        toxicology, 29
        trace analysis, 29
    crime scene preservation, 8–14
    evidence handling, 14–19
        classes, 19
        packaging and preservation, 17–19
        recognition and collection, 14–16
    evolution of practice, 5–7
Forensic Science Service Providers
        (FSSPs), 189
Forensic serology, 163
Forgery analysis, 33
FQS, *see* Forensic Quality Services
Franklin, Rosalind, 109
Friction ridge analysis, 29
Frye v. United States (1923), 61
FSSPs, *see* Forensic Science Service
        Providers
Fully continuous probabilistic genotyping
        software program, 183

Gain-of-function mutations, 138
Galton, Francis, 4
Gamma hydroxybutyrate (GHB), 25
Gas chromatography, 20, 34
Gel electrophoresis, 198–202
Genetics, 123–140
    analyzer, 72
    cytogenetics, 137–138
        chromosomal abnormalities, 138
        mitotic and meiotic, 137–138

Index

disease, 138–140
Mendelian, 124–130
  human pedigrees, 127–130
  rules of inheritance, 126–127
non-Mendelian, 130–137
  mitochondrial DNA, 133–137
  Y-chromosomal inheritance, 131–133
reproduction, 72
Gene transfer, 123
Genotypes, 126, 127, 142, 143, 145, 159
GHB, *see* Gamma hydroxybutyrate
Globally Harmonized System of Classification and Labeling of Chemicals (GHSCLC), 56
Glycophorin A (GPA), 85
Goddard, Calvin, 4–5
"Goldilocks Principle," 170
Goldman, Ronald, 9
GPA, *see* Glycophorin A
Gross, Hans, 3
Guaifenesin, 26
Guanine, 107, 109
Gunshot residue detection, 35

Hallucinogens, 22–24
Handwriting analysis, 33
Haploid cells, 123
Haptens, 161
Hardy, G. H., 141
Hardy–Weinberg principle, 142–145
Hazard communication plan, 57
Hazardous waste, 58
Hematin assay, 85–87
Hematoxylin, *see* Christmas tree stain
Hemochromogen assay, 87
Hemoglobin (Hb), 80, 85–87, 165
Heredity, 126
Heroin, 22, 25
Heterochromatin, 106
Heteroplasmy, 135
Heterozygosity, 142, 143
Homologous recombination, 120
Homozygosity, 142–143
Hot start PCR, 172
HPA, *see* Human pancreatic α-amylase
HSA, *see* Human salivary α-amylase
Human Genome Project, 167
Human pancreatic α-amylase (HPA), 93
Human salivary α-amylase (HSA), 93
Hybridoma cells, 162
Hydrocarbons, 66–67

IAAI, *see* International Association of Arson Investigators
IACIS, *see* International Association of Computer Investigative Specialists
IAI, *see* International Association for Identification
IED, *see* Improvised explosive device
IEF, *see* Isoelectric focusing
Immunoassay techniques, 163
Immunochromatographic tests, 85, 90, 92, 93, 95, 96
Immunodiffusion, 93
Immunoelectrophoresis, 93
Immunogens, 161
Immunoglobulins, *see* Antibodies
Improvised explosive device (IED), 14
Individual evidence, 19
Inductive reasoning, 76
Infrared light source, 79
Infrared spectroscopy, 20
Inheritance principles, 72
Intercalating dye assay, 210–211
Internal proficiency tests, 55
Internal validation, 52–53, 184
International Association for Identification (IAI), 42
International Association of Arson Investigators (IAAI), 42
International Association of Computer Investigative Specialists (IACIS), 50
International Organization for Standardization (ISO), 40
International Society of Forensic Genetics (ISFG), 214
International Union of Pure and Applied Chemistry (IUPAC), 66
Iodine fuming, 31
Ionic bond, 69
ISFG, *see* International Society of Forensic Genetics
ISO, *see* International Organization for Standardization
ISO 17025 standard, 41
Isoelectric focusing (IEF), 165
Isoelectric points (pI), 165
Isoenzymes, 164
IUPAC, *see* International Union of Pure and Applied Chemistry

Jaffe color test, 96

222 Index

Karyotype, 107
Keratinocytes, 135
Kinship relationships, 158–159
Kirk, Paul L., 5
Kumho Tire Co. v. Carmichael (1999), 62

LAB, *see* Laboratory Accreditation Board
Laboratory Accreditation Board (LAB), 40
Laboratory information management
    systems (LIMS), 56
Landsteiner, Karl, 5, 81
Laser capture microdissection (LCM)
    machine, 91
Latent fingerprints, 29–30
Lattes, Leone, 4
Lattes crust test, 4, 81
Law of independent assortment, 126
Law of segregation, 126
LCM, *see* Laser capture microdissection
    machine
LDIS, *see* Local DNA Index System
'Leading and lagging' strand, 114, 116
Legal decisions, 61–62
Legal definitions, 62–63
Leukocytes, *see* White blood cells
Likelihood ratio (LR), 146, 159, 192–193
Limit of detection (LoD), 53
Limit of quantitation (LoQ), 53
LIMS, *see* Laboratory information
    management systems
Linear sequential unmasking, 187
Liquid handling robots, 182
Local DNA Index System (LDIS), 160
Locard, Edmond, 3
Locard's Exchange Principle, 3, 4
LoD, *see* Limit of detection
London dispersion forces, *see* Dipole-dipole
    interaction
Loop pattern of fingerprint, 32
LoQ, *see* Limit of quantitation
Loss-of-function mutations, 138
LR, *see* Likelihood ratio
LR Mix Studio, 182
LSD, *see* Lysergic acid diethylamide
Luminol, 84
Lysergic acid diethylamide (LSD), 23, 25

Male-specific Y region (MSY), 131, 132
Marijuana, 22, 25
Massively parallel sequencing (MPS), 139
Mass spectrometry, 20, 34
McCrone, Walter C., 5

MDMA (Ecstasy/Molly), 23, 25
Meiosis, 108, 123, 127, 128, 137–138
Melanocytes, 135
Melanosome, 135
Melendez-Diaz v. Massachusetts (2009), 62
Mendel, Gregor, 72, 124–125
Messenger RNA (mRNA), 112, 178
Metagenomics, 152
Methadone, 22
Methamphetamines, 25
Methaqualone, 25
MHS-5, *see* Mouse anti-Human-Sperm
    antibody number 5
Microbial forensics, 152
Microsatellites, 178, 179
Microspectrophotometer, 197
Minutiae, 32
Mitochondria, 101
Mitochondrial DNA (mtDNA), 108, 130,
    133–137, 158
Mitosis, 108–109
Monoclonal antibodies, 161
Morphine, 22
Mouse anti-Human-Sperm antibody
    number 5 (MHS-5), 92
MPS, *see* Massively parallel sequencing
mRNA, *see* Messenger RNA
MSY, *see* Male-specific Y region
mtDNA, *see* Mitochondrial DNA
Mullis, Kary, 168
MUP, *see* 4-methylumbelliferone phosphate
Mutation mechanisms and rates, 118–121
    kinds, 118–119
        insertions/deletions, 118–119
        rare, 119
        substitutions and frameshifts, 118
        unequal crossing over, 118
        unstable trinucleotide repeats, 118
    repair, 119

Narcotics, 21
National Academy of Sciences (NAS), 50,
    73, 147
National Commission on Forensic Science
    (NCFS), 50, 188, 190
National DNA Index System (NDIS), 160
National Institute for Standards and
    Technology (NIST), 51,
    54–55, 148
National Research Council (NRC), 147
NCFS, *see* National Commission on
    Forensic Science

Index

223

NDIS, *see* National DNA Index System
Next-generation sequencing (NGS), 139
NFR, *see* Nuclear Fast Red
NGS, *see* Next-generation sequencing
Ninhydrin spray, 31
NIST, *see* National Institute for Standards and Technology
NIST Policy on Metrological Traceability, 54
Nondisjunction, 137–138
Nonhomologous end joining, 120
Non-human molecular application, 149–152
  animal forensic DNA, 150
  microbial DNA, 152
  plant forensic DNA, 150–152
Nonrecombining region (NRY), 132
NRC, *see* National Research Council
NRY, *see* Nonrecombining region
Nuclear Fast Red (NFR), 91
Nucleic acids isolation and purification, 205–209
  Chelex extraction, 208–209
  differential extraction, 207–208
  organic extraction, 206–207
  silica-based extraction, 209
Nucleotides, 109

O blood, 81
Occupational Safety and Health Administration (OSHA), 57
Okazaki fragments, 114
Oligo (dT) primers, 178
Oligospermia, 89
Opiates, 21
Orfila, Mathieu, 2
Organic extraction, 206–207
Organization of Scientific Area Committees (OSAC), 51
Osborn, Albert S., 4
OSHA, *see* Occupational Safety and Health Administration
Ouchterlony test, 87
Oxycodone, 22

p30, *see* Prostate-specific antigen
PAR, *see* Pseudoautosomal region
Patau syndrome, 138
Patent fingerprints, 30
Paternity Index (PI), 159
Pathological examinations, 6
PCR, *see* Polymerase chain reaction
p-dimethylaminocinnamaldehyde (DMAC), 96

Personal protective equipment, 10
Personnel certification, 41–43, 50
PGM, *see* Phosphoglucomutase
Phadebas reagent, 93, 94
Phase contrast microscope, 196, 197
Phenotypes, 126, 129, 130, 145
Phosphoglucomutase (PGM), 165
Photographic evidence, 12
Physical evidence, 155
Physiological fluids biochemistry, 99
pI, *see* Isoelectric points
PI, *see* Paternity Index
Picroindigocarmine (PIC), 91
Pitchfork, Colin, 167
Plant cells, 71
Plasma, 79–80
Platelets, 80
Polarized microscopes, 196
Polyacrylamide gel, 199, 201
Polyclonal antibodies, 161
Polymerase chain reaction (PCR), 54, 150, 151, 168–170, 172–178, 182, 203, 207, 211–212
Population genetics, 141–148
  Hardy–Weinberg, 142–145
  mechanisms of evolution, 145–146
    mutation, 145
    selection, 145–146
  population databases, 148
  statistics and probability, 146–147
    likelihood ratios, 146
    Pd and Pi, 147
Precipitin test, 150
Preventative action, 55
Primary binding reactions, 163
Primer dimers, 173
Primers, 171–174
Probability, 146
  genotyping software, 182
  interpretation, 75
Procedural law, 64–65
Proficiency tests, 55–56
Prokaryotic cells, 99–100
Prostate-specific antigen (PSA), 91–93
Protein and enzyme polymorphisms, 161, 164–165
PSA, *see* Prostate-specific antigen
Pseudoautosomal region (PAR), 131–132
Psilocin, 23
Psilocybin, 23
Punnett square, 127, 129

224 Index

QAS, *see* Quality assurance standards
qPCR, *see* Quantitative PCR
Quality Assurance Manager, 52
Quality assurance standards (QAS), 40, 41, 58, 63, 191–192
    for DNA databasing laboratories, 183
    for forensic DNA laboratories, 183
Quality control and quality assurance, 39–58
    accreditation, 39–40
    accrediting bodies, 40–41
    application, 51–56
    document/data management, 56
    personnel certification, 41–43, 50
    safety, 56–58
        chemical hygiene, 56–57
        hazardous waste, 58
        universal precautions, 58
    standardization, 50–51
Quantitative PCR (qPCR), 170–171, 211
Questioned documents, 33

Random hexamer primers, 178
Random match probability (RMP), 192
Rapid Stain Identification (RSID), 85
Ras gene, 138
rCRS, *see* Revised Cambridge Reference Sequence
Real time PCR (RT-PCR), 87, 93, 95, 97, 170–172, 203
Reconstructed Sapiens Reference Sequence (RSRS), 137
Red blood cells, 79–80, 164
Relative fluorescence units (RFU), 203, 213
Restriction endonucleases, 167
Restriction fragment length polymorphism (RFLP), 166–167, 179, 207
Retrotransposons, 180
Reverse transcriptase PCR (RT-PCR), 178
Revised Cambridge Reference Sequence (rCRS), 136
RFLP, *see* Restriction fragment length polymorphism
RFU, *see* Relative fluorescence units
Rhesus (Rh) factor, 5–6, 81
Ribosome, 112
Rifling marks, 34
RMP, *see* Random match probability
RNA, 111–112, 133
Roofies (Rohypnol), 25
Rrestriction enzymes, *see* Restriction endonucleases

RSID, *see* Rapid Stain Identification
RSID™-Saliva kit, 94
RSID™-Urine, 96
RSRS, *see* Reconstructed Sapiens Reference Sequence
RT-PCR, *see* Real time PCR; Reverse transcriptase PCR

Safety, 56–58
    chemical hygiene, 56–57
    hazardous waste, 58
    universal precautions, 58
Safety data sheets (SDS), 56
SALIgAE®, 94
Saliva, 93–95
    confirmatory tests, 94–95
    presumptive tests, 93–94
Satellite DNA, 178
Saturated hydrocarbons, 67
Scanning electron microscope (SEM), 198
Schlesinger test, 97
Science terms and principles, 65–78
    biology, 71–72
    chemistry, 66–71
    logic, 75–76
    physics, 72–73
    physiology, 73
    statistics, 73–75
Scientific Working Group on DNA Analysis Methods (SWGDAM), 41, 53, 133, 183–184, 214, 215
Scientific working groups (SWGs), 51
Screening tests, 20, 79
SDIS, *see* State DNA Index System
SDS, *see* Safety data sheets
Secondary binding reactions, 163
SEM, *see* Scanning electron microscope
Semen, 88–93
    components, 88–90
    confirmatory testing, 91–93
    pH, 99
    presumptive tests, 89–91
Semenogelin, 91–92
Semi-continuous probabilistic genotyping software program, 182
Seminal fluid, 88, 90
Seminal vesicle-specific antigen (SVSA), 91, 92
SERATEC® HemDirect, 84
Serology, 20, 79, 81
Serum, 80–81
Short tandem repeats (STRs), 168

Index

225

fragment analysis, 212–215
   application/processes, 213
   interpretation/results, 213–215
   theory, 212–213
   testing, 136, 179
Silica-based extraction, 209
Simple sequence repeats (STRs), *see*
   Microsatellites
Simpson, Nicole Brown, 8
Single nucleotide polymorphisms (SNPs),
   138, 139, 178
Slot blot assay, 210
SNPs, *see* Single nucleotide polymorphisms
Solvent-based extraction, *see* Organic
   extraction
Southern transfer and hybridization, 167
Spectrophotometer, 197
Spermatozoa, 88, 91
Sperm cells, 88–89, 91, 103, 207–208
Spermine, 90
Standard deviation, 74–75
Starch-iodine assays, 93
State DNA Index System (SDIS), 160
Statistical analysis, 146
Stereoisomers, 70
Stereoscopic microscope, 196
StereoZoom, *see* Stereoscopic microscope
Stimulants, 25
"Strengthening Forensic Science in
   the United States: A Path
   Forward," 50
STRmix, 182, 183
STRs, *see* Short tandem repeats
Subpoena, 62–63, 65
Superglue, *see* Cyanoacrylate fuming
SVSA, *see* Seminal vesicle-specific antigen
SWGDAM, *see* Scientific Working Group
   on DNA Analysis Methods
SWGs, *see* Scientific working groups

Takayama crystal assay, *see*
   Hemochromogen assay
Tamm-Horsfall protein (THP), 96
Taq polymerase, 171–173, 176
Technical working groups (TWGs), 51
Teichmann crystal assay, *see* Hematin assay
Telomeres, 103
Tertiary binding, 163
Tetrahydrocannabinol (THC), 22
Tetramethylbenzidine (TMB), 210
THC, *see* Tetrahydrocannabinol
Thermal energy, 72

THP, *see* Tamm-Horsfall protein
Thymine, 107, 109
TMB, *see* Tetramethylbenzidine
Tool marks, 36, 43
Toxicology, 6, 29
Tranquilizers, 24, 26
Transcription and translation, 111–112
Transfer RNA (tRNA), 112
Tris-HCl, 205
Trisomy, 137, 138
tRNA, *see* Transfer RNA (tRNA)
TrueAllele, 182–183
TWGs, *see* Technical working groups

United Nations Office on Drugs and Crime
   (UNODC), 50
United States Postal Service lab, 20
UNODC, *see* United Nations Office on
   Drugs and Crime
Unsaturated hydrocarbons, 67
Urine, 95–97
   confirmatory tests, 96–97
   presumptive tests, 96
   properties, 95–96
Uritrace, 96
Urobilinoids test, 97
Uromodulin, *see* Tamm-Horsfall protein

Validation Guidelines for DNA Analysis
   Methods, 184
Valium, 26
Van der Waals forces, 70
Variable number of tandem repeats
   (VNTR), 167, 178–179
Video recording, 11
Visual techniques, 79
VNTR, *see* Variable number of tandem
   repeats
Voir dire, 62
Vollmer, August, 5

Warkany syndrome 2, 138
Watson, James, 109
Weinberg, Wilhelm, 141
White blood cells, 79–80
Whorl pattern of fingerprint, 32

X-ray diffraction, 109

Y chromosome, 127, 128, 131, 132, 133
Y-Short Tandem Repeat (Y-STR) DNA
   testing, 130, 132–133, 135, 188

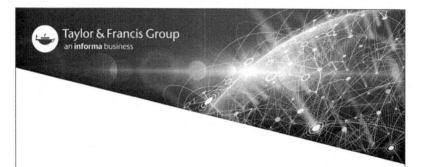